T0143375

From Artificial Intelligence to Brain Intelligence

Rajiv Joshi

IBM Research Division, USA

Matthew Ziegler

IBM Research Division, USA

Arvind Kumar

IBM Research Division, USA

Eduard Alarcon

Technical University of Catalunya,
UPC BarcelonaTech, Spain

Tutorials in Circuits and Systems

For a list of other books in this series, visit www.riverpublishers.com

Series Editors

Peter (Yong) Lian

President IEEE
Circuits and Systems Society
York University, Canada

Franco Maloberti

Past President IEEE
Circuits and Systems Society
University of Pavia, Italy

Published, sold and distributed by:
River Publishers
Alsbjergvej 10
9260 Gistrup
Denmark
www.riverpublishers.com

Availability: February 2020

ISBN: Print: 9788770221238
     E-book: 9788770221245

Copyright © 2020 by The Institute of Electrical and Electronics Engineers - Circuits and Systems Society (IEEE-CASS)

Published by River Publishers.

No part of this publication may be reproduced, stored in are retrieval system, or transmitted in any form or by any means, electronic, mechanical, photocopying, recording, scanning, or otherwise, except as permitted under Section 107 or 108 of the 1976 United States Copyright Act, without either the prior written permission of the IEEE-CASS. Requests to the IEEE-CASS for permission to reuse content should be addressed to the IEEE-CASS Intellectual Property Rights Office at manager@ieee-cas.org

Library of Congress Cataloging-in-Publication Data: February 2020
Editors: Rajiv Joshi, Matt Ziegler, Arvind Kumar y Eduard Alarcon
Title: From Artificial Intelligence to Brain Intelligence
Sub-title: AI Compute Symposium 2018

Table of contents

Introduction

The field of AI is not new to researchers, as its foundations were established in the 1940s. After many decades of inattention, there has been a dramatic resurgence of interest in AI, fueled by a confluence of several factors. The benefits of decades of Dennard scaling and Moore's law miniaturization, coupled with the rise of highly distributed processing, have led to massively parallel systems well suited for handling big data. The widespread availability of big data, necessary for training AI algorithms, is another important factor. Finally, the greatly increased compute power and memory bandwidths have enabled deeper networks and new algorithms capable of accuracy rivaling that of human perception. Already AI has shown success in many diverse areas, including finance (portfolio management, investment strategies), marketing, health care, transportation, gaming, defense, robotics, computer vision, education, search engines, online assistants, image/facial recognition, anomaly detection, spam filtering, online customer service, biometric sensors, and predictive maintenance, to name a few. Despite these remarkable advances, the human brain is still superior in many ways – including, notably, energy efficiency and one-shot learning – giving researchers new areas to explore. In summary, AI research and applications will continue with vigor in software, algorithms, and hardware accelerators. These exciting developments have also brought new questions of ethics and privacy, areas which must be studied in tandem with technological advances. To continue the success story of AI the AI Compute symposium was launched with the sponsorship of IBM, IEEE CAS and EDS for the first time. The aim of this publication is to compile all the materials presented by the renowned speakers in the symposium into a book format, serving as a learning tool for the audience.

Together with the IEEE Circuits and Systems Society (CAS) and the IEEE Electron Device Society (EDS), IBM Research led the 1st AI Compute Symposium at the IBM T.J. Watson Research Center THINKLab in Yorktown Heights, NY, on October 25th 2018. This symposium brought dreamers, thinkers, and innovators across industry and academia together for a one-day symposium focusing on cutting-edge research addressing AI Compute challenges and future directions of AI. The symposium consisted of two keynotes, six invited talks, a student poster session, and a panel discussion. The event was free of charge and had over 155 attendees from IBM, various companies and universities. IBM along with IEEE indeed showcased leadership and advancement in AI-Compute domain.

Keynote talks were delivered by Lisa Amini from IBM and Rob Aitken from ARM. Lisa Amini provided an inspiring overview of research projects from the MIT-IBM Watson AI Lab, which has recently celebrated a one-year anniversary. Amini described three tiers of AI research spanning narrow, broad, and general AI. She posited that the AI research community is beginning a journey into broad AI, whereas general AI is still a long-term goal for the future. Rob Aitken followed with a keynote address describing how many emerging AI problems present dynamically changing goals and rules, rather than the fixed goals and rules of conventional computing problems. Aikten also presented practical approaches for decomposing complex problems into manageable components that may provide a path for tackling complex AI challenges. Finally he concluded that the IoT needs distributed AI systems, with AI and ML application to IoT involves with real time processing, explainability and security as key challenges.

Following the keynotes, Mike Davies from Intel and Jeff Burns from IBM gave invited talks during the "Industry Perspectives" session. These talks provided a landscape of industrial research for both near and long term, spanning architectural, circuit design, and semiconductor technology. Mike Davies' talk focused on Intel's Loihi neuromorphic chip as well as future directions in neuromorphic research. Although Loihi is a digital chip, this avenue of research pushes beyond conventional von-Neumann architectures. On the other hand, Jeff Burns' talk focused on current efforts and future plans for deep learning acceleration. Burns described a vision beginning with specialized digital accelerators in the near term with enhancements based on analog circuit design and future device technology in the future.

Next, in the "Bio-inspired Computing" session, Andreas Andreou from Johns Hopkins University provided a number of examples of bio-inspired chip designs, many of which are components in systems that solve complex problems of interest to organizations like DARPA. Inarguably the most provocative talk of the day, Todd Hylton from the University of California, San Diego, proposed the concept of thermodynamic computing as a potential future direction for computing research. Its evolution can be biased through programming, training and rewarding.

The third session on "Emerging Technologies" included talks by Wei Lu from the University of Michigan and Naveen Verma from Princeton University. Lu described recent research progress on RRAM (resistive random access memory) device and chip-level design and fabrication. He described how RRAM can provide a platform for neuromorphic computing, which is a promising direction for future AI computing. Naveen Verma delivered a case for circuit and architectural approaches for in-memory computing, another topic of high interest to the community. He presented measurement results from several fabricated chips providing compelling evidence for in-memory computing potential.

The symposium also had a well-attended student poster session, where about 30 students presented compelling research spanning numerous topics in AI computing. Two best poster presentations were awarded. One award was given to Sohum Datta from UC Berkeley for work on "A 2048-dim General-purpose Hyper-Dimensional Processor." A second award was given to Jingcheng Wang from the University of Michigan for "Neural

Cache: Bit-Serial In-Cache Acceleration of Deep Neural Networks."

The symposium closed with a panel discussion entitled "Artificial Intelligence or Artificial Stupidity: How smart will smart AI be?". The panelists included the keynote and invited speakers, as well as IBM Fellow Mark Wegman. A lively and at times heated debate ensued where topics from the progress in AI research to AI ethics were touched upon. Todd Hylton presented a case that while narrow AI challenges have made progress, the community is far from approaching true intelligence. At times Andreas Andreou and Mark Wegman sparred over the future of AI research progress, while Naveen Verma wound up as the mediating voice. The panel discussion is scheduled to appear IEEE TV in the near future.

Overall, the general consensus of attendees, speakers, and organizers was that day provided a great platform for educational forum and lively discussions related to the most current compelling topics in the computing field. Additional publications (Book, Journal papers etc) based on the symposium technical content are planned to provide educational resources for anyone interested. Although early in the stages, future events based on AI Compute are being planned by IBM and IEEE. Please see the following website for updates.

http://ibm.biz/AIcomputesymposium

CHAPTER 01

Research Directions in AI Algorithms and Systems

Lisa Amini

IBM

This first chapter is an inspiring overview of research projects from the MIT-IBM Watson AI Lab, which has recently celebrated a two-year anniversary. Lisa Amini describes three tiers of AI research spanning narrow, broad, and general AI. Currently we are practicing narrow AI, meaning performing only a single task in a single domain. Recommending an item from a list, spam filtering tools, and image recognition are examples of narrow AI. These narrow AI approaches are excellent in executing a focused complex task, but their abilities are limited in other domains. On the other hand, general AI involves unsupervised and autonomous systems. Drones guiding each other without any supervision and performing several tasks is an example of general AI. Today, the AI research community is beginning a journey into broad AI, whereas general AI is still a long-term goal for the future. The chapter also provides a sampling of IBM-MIT joint work in the areas of AI algorithms, physics of AI, ethics of AI, and AI for social good.

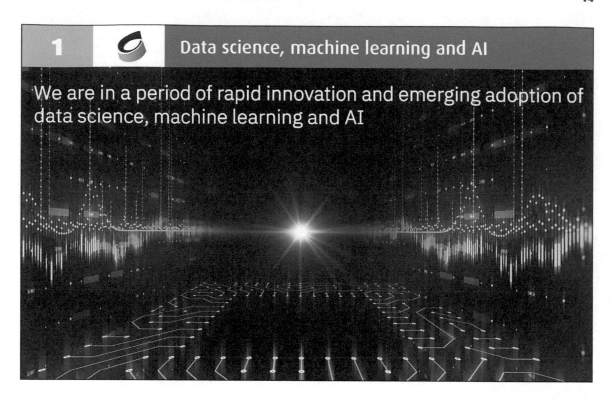

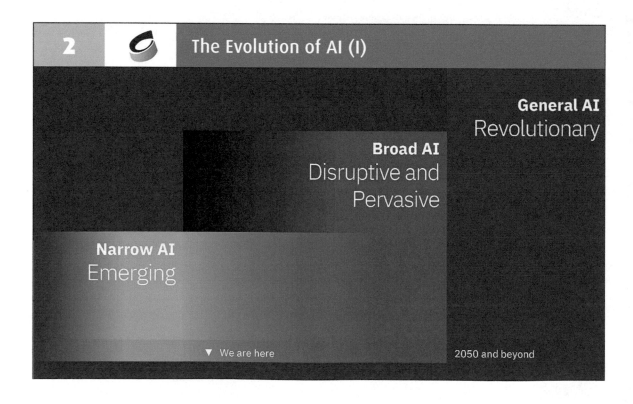

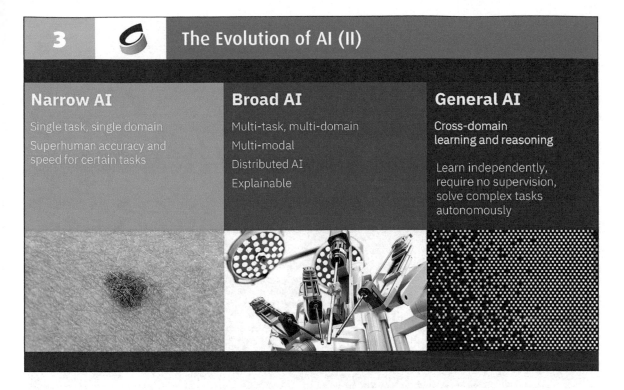

3 — The Evolution of AI (II)

Narrow AI
Single task, single domain
Superhuman accuracy and speed for certain tasks

Broad AI
Multi-task, multi-domain
Multi-modal
Distributed AI
Explainable

General AI
Cross-domain learning and reasoning

Learn independently, require no supervision, solve complex tasks autonomously

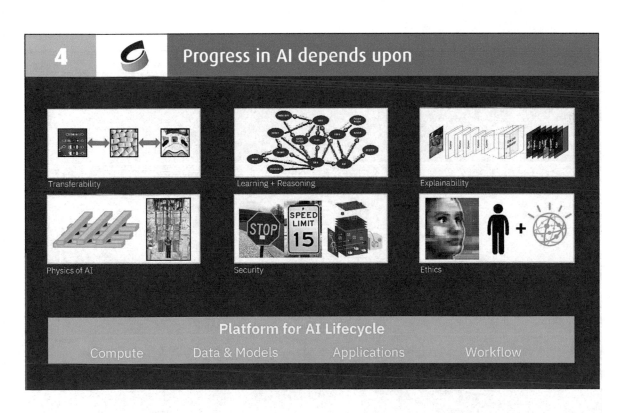

4 — Progress in AI depends upon

Transferability

Learning + Reasoning

Explainability

Physics of AI

Security

Ethics

Platform for AI Lifecycle

Compute   Data & Models   Applications   Workflow

5 · MIT-IBM Watson AI Lab

$240M 10 year commitment to jointly create the future of artificial intelligence

AI algorithms

- Learning and Reasoning

- Continuous, multi-task, small data, explanations, ...

Physics of AI

- Analog AI

- AI & Quantum

http://mitibmwatsonailab.mit.edu/

Applications of AI to industries

- Healthcare, Life Sciences

- Cybersecurity

Advancing shared prosperity through AI

- Ethics of AI, fair, unbiased

- AI for Social Good

6 · Better leveraging available data

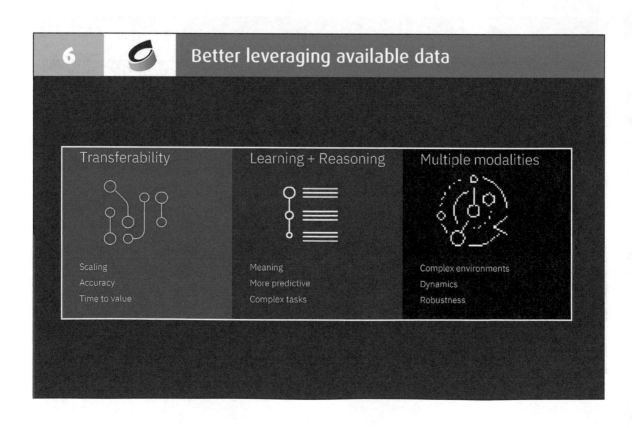

Transferability

Scaling
Accuracy
Time to value

Learning + Reasoning

Meaning
More predictive
Complex tasks

Multiple modalities

Complex environments
Dynamics
Robustness

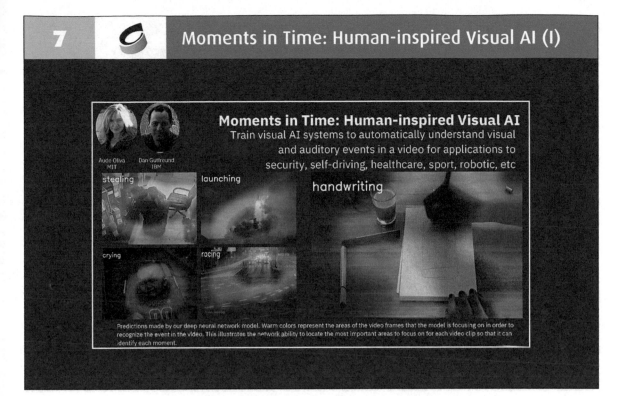

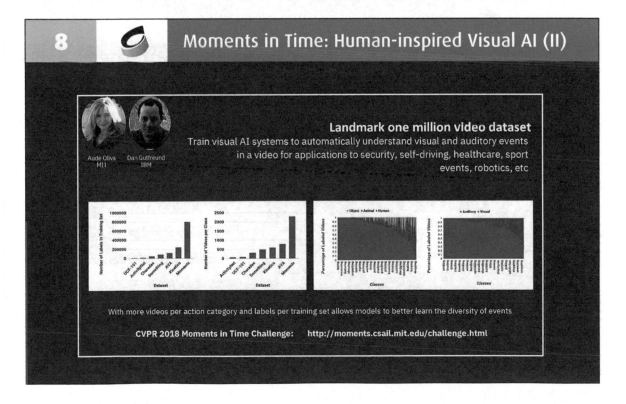

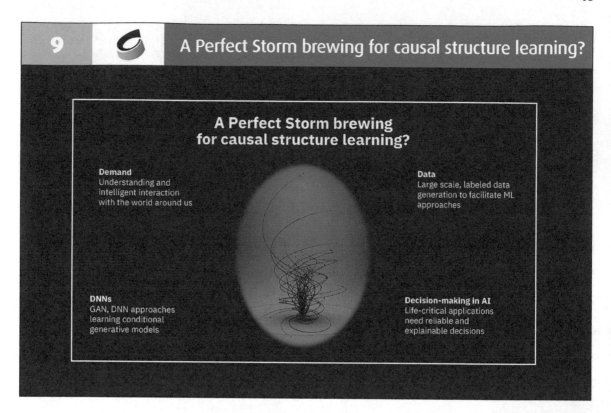

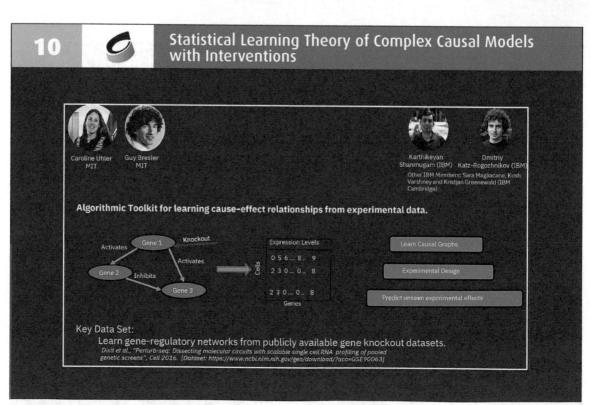

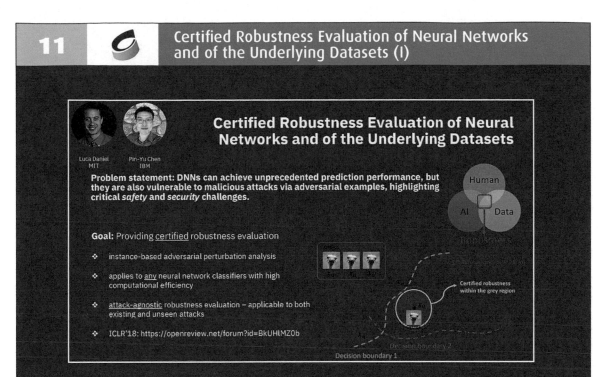

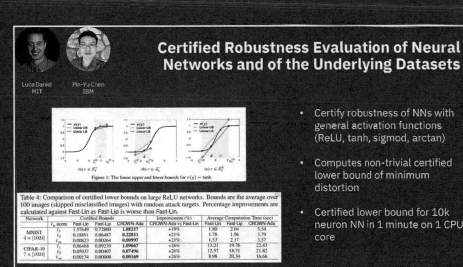

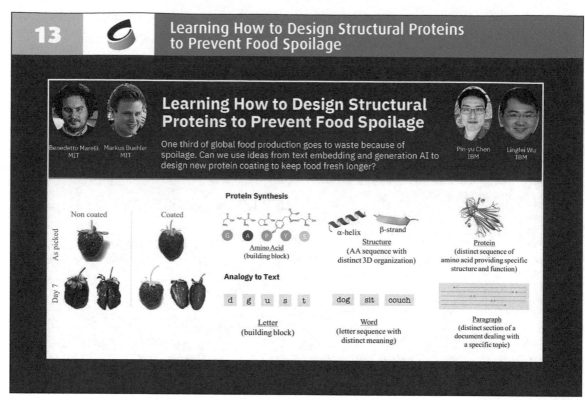

13 — Learning How to Design Structural Proteins to Prevent Food Spoilage

Learning How to Design Structural Proteins to Prevent Food Spoilage

Benedetto Marelli — MIT
Markus Buehler — MIT
Pin-yu Chen — IBM
Lingfei Wu — IBM

One third of global food production goes to waste because of spoilage. Can we use ideas from text embedding and generation AI to design new protein coating to keep food fresh longer?

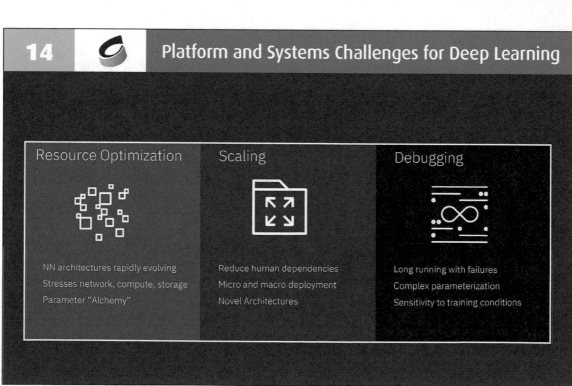

14 — Platform and Systems Challenges for Deep Learning

Resource Optimization

NN architectures rapidly evolving
Stresses network, compute, storage
Parameter "Alchemy"

Scaling

Reduce human dependencies
Micro and macro deployment
Novel Architectures

Debugging

Long running with failures
Complex parameterization
Sensitivity to training conditions

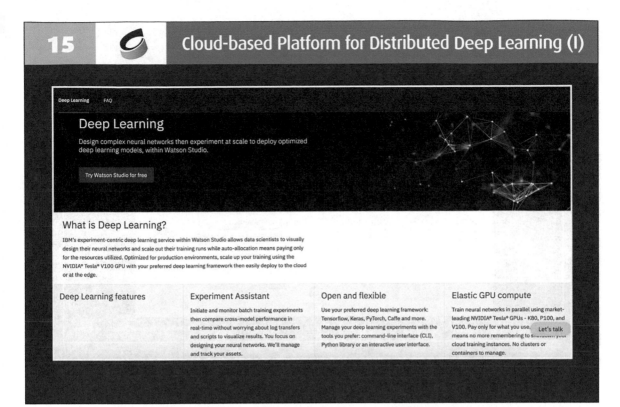

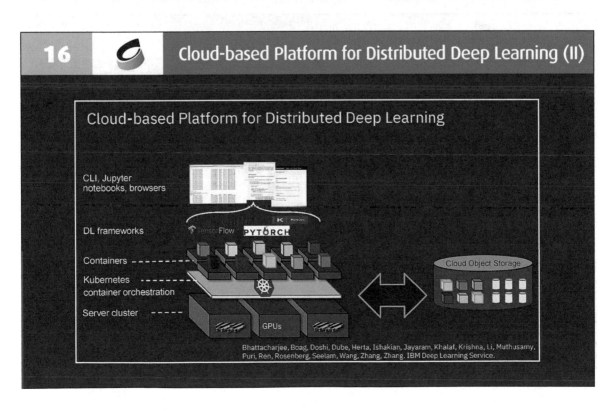

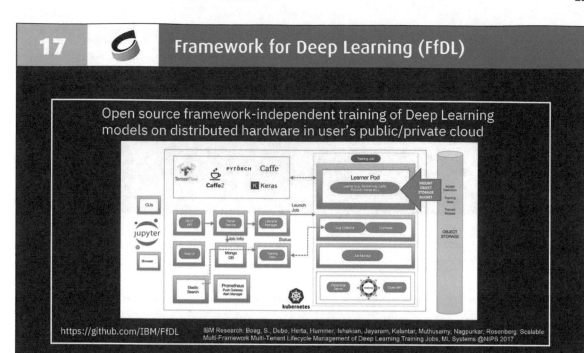

17 — Framework for Deep Learning (FfDL)

Open source framework-independent training of Deep Learning models on distributed hardware in user's public/private cloud

https://github.com/IBM/FfDL

IBM Research: Boag, S., Dube, Herta, Hummer, Ishakian, Jayaram, Kalantar, Muthusamy, Nagpurkar, Rosenberg. Scalable Multi-Framework Multi-Tenant Lifecycle Management of Deep Learning Training Jobs, ML Systems @NIPS 2017

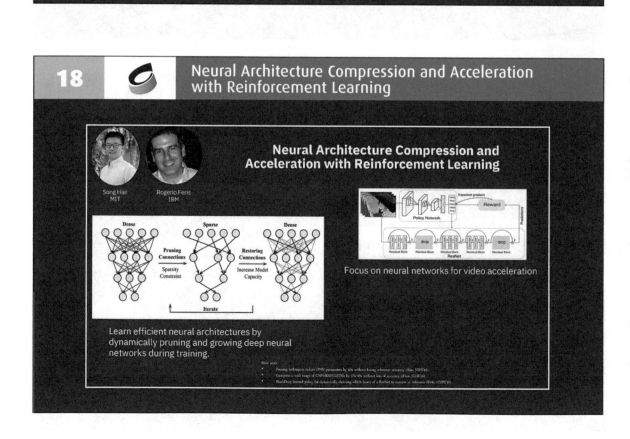

18 — Neural Architecture Compression and Acceleration with Reinforcement Learning

Neural Architecture Compression and Acceleration with Reinforcement Learning

Song Han
MIT

Rogerio Feris
IBM

Focus on neural networks for video acceleration

Learn efficient neural architectures by dynamically pruning and growing deep neural networks during training.

19　Quantifying Information Flows in Deep Neural Networks (I)

Yury Polyanskiy
MIT

Brian Kingsbury
IBM

Quantifying Information Flows in Deep Neural Networks

- How does information about a DNN's input evolves with its processing by successive layers?
- Quantification via Mutual Information results in a hard estimation problem.

- Approach:
  - Study networks with internal noise so the Mutual Information between the input X and an internal representation T is meaningful
  - Develop Mutual Information estimator for noisy networks with theoretical convergence rate guarantees

$$T_{\ell-1} \rightarrow \phi\left(W_\ell^{(k)} T_{\ell-1} + b_\ell(k)\right) \xrightarrow{S_\ell(k)} \oplus \xrightarrow{T_\ell(k)}$$
$$Z_\ell(k) \sim \mathcal{N}(0, \beta^2)$$

20　Quantifying Information Flows in Deep Neural Networks (II)

Yury Polyanskiy
MIT

Brian Kingsbury
IBM

Quantifying Information Flows in Deep Neural Networks

Fully Connected Network with tanh nonlinearities, 3-digit MNIST Classifier

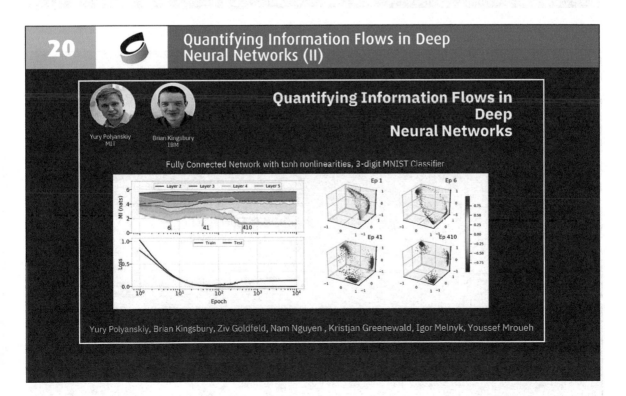

Yury Polyanskiy, Brian Kingsbury, Ziv Goldfeld, Nam Nguyen , Kristjan Greenewald, Igor Melnyk, Youssef Mroueh

21 **Errors in Artificial Neural Networks**

Errors in Artificial Neural Networks
Towards Debugging AI
Team: Dan Gutfreund (IBM), David Bau (MIT), Evan Phibbs (IBM)

Antonio Torralba
MIT

Stefanie Jegelka
MIT

Hendrik Strobelt
IBM

"Debugging of Neural Networks" is a major challenge in AI. The project investigates how to **classify and detect** the types of errors (bugs) that neural networks make, how to **find their causes**, and how **human intervention** can help mitigating them.

Network output: Washing dishes

Ground truth: Brushing teeth

| unit 1679: Bathroom | unit 867: Kitchen | unit 1749: House | unit 795: Bathroom | unit 1978: Person |

Conclusion: the network seems confused about the scene. It did not detect the brush.

---

22 **Machine Learning in Hilbert Space**

Machine Learning in Hilbert Space
Using quantum computers for machine learning

Aram Harrow
MIT

Peter Shor
MIT

Sergey Bravyi
IBM

Kristan Temme
IBM

Small data sets, sophisticated analyses

Large data sets, simpler analyses

Classical input data
101100011010
111010101000
100101010101

Quantum state generation

Encodes superposition of exponentially many data attributes

Quantum postprocessing

Classical output
010100010
010101000
101011110

Classical computer

101100011010

Quantum Computer

010010010100

Apply ML techniques to classical representations of quantum data to improve simulation methods for quantum circuits

Theoretical analysis targeting quantum advantage for specific computation problems in AI

23 More Resources for Systems Researchers in Deep Learning

Watson Machine Learning
- Cloud access to ML, DL, Jupyter notebooks, ...
- http://www.ibm.com/cloud/machine-learning

Fabric for Deep Learning (FfDL)
- Framework-independent training of Deep Learning models on distributed hardware in user's public/private cloud
- http://github.com/IBM/FfDL

Adversarial Robustness Toolkit
- Rapid crafting and analysis of attacks and defense methods for machine learning models
- http://github.com/IBM/adversarial-robustness-toolbox

CHAPTER 02

An ARM Perspective on Hardware Requirements and Challenges for AI

Rob Aitken

Arm Research

This chapter by Rob Aitken describes how many emerging AI problems present dynamically changing goals and rules, rather than the fixed goals and rules of conventional computing problems. Practical approaches for tackling AI challenges are suggested based on decomposing complex problems into manageable components. The chapter focuses mainly on AI for edge and IoT (internet-of-things).

Distributed heterogeneous systems provide high potential for AI processing within an IoT paradigm. These emerging distributed systems spread the compute tasks across various devices, as opposed to today's centralized datacenter-centric AI systems. Although promising, Aitken describes three key challenges for AI at the edge: real-time processing at scale, security, and explainable AI. Next, Aikten reviews the design considerations of a recent ARM ML processor that has been designed with future IoT systems in mind. The use case of keyword spotting is overviewed as an example IoT application requiring AI processing.

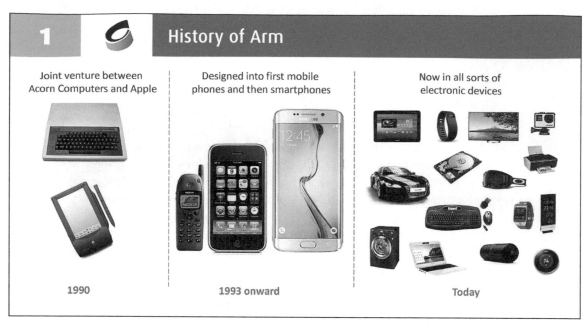

A RM has expanded dramatically from its origins as a processor developed for Acorn to its current role as the world's leading provider of semiconductor IP.

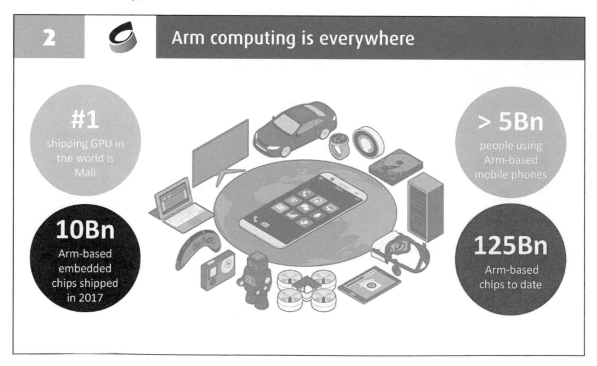

C omputing has now become a central part of our everyday life thanks to the continuous evolutions of the smartphone to become our primary compute platform.

Smartphones have enabled computing everywhere from smallest devices such as smartwatches and VR goggles to smart homes and smart cars.

ARM compute technology is deployed throughout the ecosystem - beyond mobile, from sensor to cloud.

PART 1   AN ARM PERSPECTIVE ON ~~HARDWARE REQUIREMENTS AND~~ CHALLENGES FOR AI

L et's start with challenges for AI and look at hardware requirements after that.

---

3   AI Challenges: Calvinball vs Go

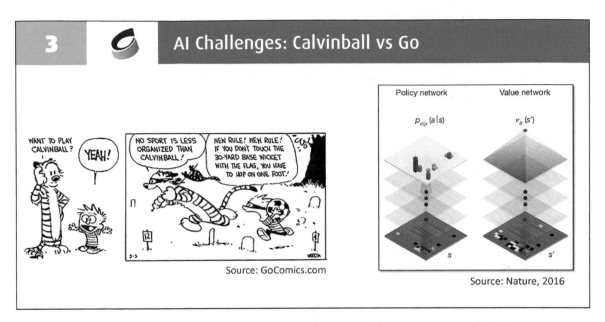

Source: GoComics.com

Source: Nature, 2016

We all know the story of AlphaGo beating humans at what was believed to be the world's most difficult game. Some of you may also remember "Calvinball" from the Calvin and Hobbes comic strip. This game featured an ever-evolving set of rules and scoring methods.

---

4   Complex problem classes

Go Problems

- Fixed goal
- Fixed rules
- Fixed interpretation of the rules
- Replay game, same results

Calvinball Problems

- Goal keeps changing
- Rules keep changing
- Interpretation of rules keeps changing
- Replay game, different results

Go and Calvinball are not only different games, but they have fundamentally different properties. Go stays fixed, so machines are able to learn how to play and eventually how to win. Calvinball, like many games played by small children, changes according to the whims of the players. Learning how to play it now does not guarantee anything about the ability to play it five minutes into the future. Current machine learning techniques are unable to handle Calvinball style problems.

5   David Marr's "Personal View" of AI, 1977

Type 1 theory ("clean"): The problem has a method to solve it; i.e. a known way of stating what the problem is and what the solution looks like

> Fourier transform, Go

Type 2 theory ("messy"): The problem does not have a type 1 theory; e.g. a problem that is solved by the simultaneous interaction of a large number of processes, whose interaction is its own simplest description

> Protein folding, Calvinball

"The principle difficulty in AI is that one can never be sure whether a problem has a type 1 theory"

David Marr came up with a similar means of classifying problems 40 years ago. There is no reason that all theories should be type 1, so there may be some problems that are inherently challenging to solve algorithmically.

6   My excellent AI adventure

I worked on artificial intelligence back in the 1980s at the Alberta Research Council in Calgary. The lab worked on various aspects of AI and robotics. I was working on an expert system. Like me, most of the lab members went on to other things, but Sheila McIlraith, now at U Toronto, remains a world authority on AI.

7 Expert systems are a type 1 solution

They are a simple way of assembling known conclusions from a set of if-then statements

Can summarize current state of knowledge on a topic

Can't invent something new

As a refresher, expert systems can interpolate answers within a fixed range of inputs and known expert responses. Humans are good at extrapolation. As an example, my grandson at one point dumped a bottle of syrup on his dog. His mother had to come up with a response to a challenge that no one could have predicted. It is extremely unlikely that any expert solution is available for this case.

8 Convolutional neural nets are a type 1 solution

Input produces defined output

Defined approximation of an arbitrary function

Very good at a variety of problems (e.g. image classification)

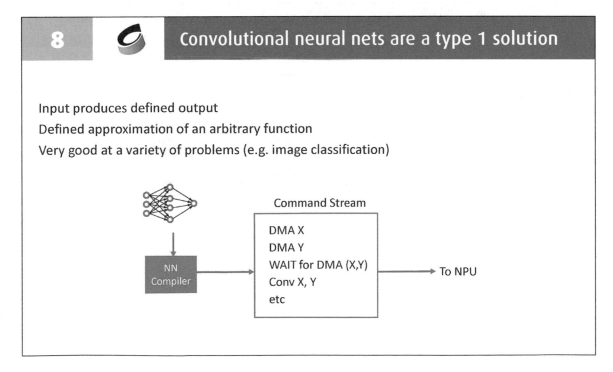

One way to look at neural nets is as highly adaptable tool for highly accurate curve fitting for arbitrary functions.

9 Humans create Calvinball problems

- What is the meaning of life?
- What is intelligence?
- When is bedtime?
- Is this program verified to be correct?

10 Things that programs can do better than humans

- Play Go
- Compile software
- Implement hardware
- Monitor security cameras

Source: New Yorker

Computers can do more every day.

11 Is driving Calvinball or Go?

Science

Google promises autonomous cars for all within five years

New California law clears driverless cars from 2015

By Iain Thomson in San Francisco 25 Sep 2012 at 23:13   108 💬   SHARE ▼

The Register, Sept 25 2012

GM's Driverless Vehicles Require a 'Degree of Harmonization' With Governments, Innovation Chief Says

Fortune, Oct 15, 2018

Chicago, Oct 14, 2017

D riving is somewhat like Calvinball: There are rules, but they have flexible interpretations.

12 Avoiding Calvinball by moving the goalposts

Create a type 1 problem
- Define (or declare) a fixed-complexity subset of the hard problem to be important
- Define a metric for success
- Build solution
- Iterate and make solution (and metric) better

N ew situations arise regularly. Most automated solutions to real life Calvinball problems involve changing the definition of success.

PART 2 AN ARM PERSPECTIVE ON HARDWARE REQUIREMENTS ~~AND~~ CHALLENGES FOR AI

N ow onto the hardware...

13   Observations

AI is more than machine learning

Machine learning is everywhere, and increasingly moving to the edge

IoT is expanding

Moore's law is slowing down

H ere's where we are at now.

14   Aside: Making predictions

"Prediction is difficult. Especially about the future"
  • Attributed to pretty much everybody you can think of

California weather prediction: same tomorrow as today
  • 90%+ accurate, but useless

"Superforecasters" – break complex problems down into smaller ones

"Always verify your Internet quotes"
-Will Shakespeare

P rediction is difficult, and making useful predictions is harder than making accurate predictions.

15 So how should we predict the future?

Scale back our ambitions

Look at the data

Focus on specific disruptions and see where they go

What is the future of AI?

Even so, we can still progress...

16 Challenge: Developing, deploying and managing secure IoT at scale

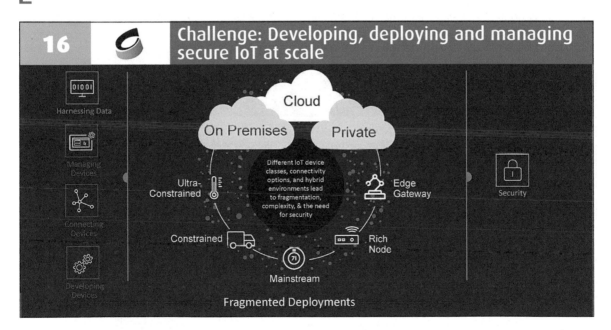

Example: Internet of things,
  Before you get to crunch the data – many other hard problems.
  - Design or procure devices that fit other constraints and inter operate.
  - Connect through various wireless protocols.
  - Maintain through cloud based management interfaces.

- Fragmentation.
- Security – not just communication level but device management too.

Then you might extract value from data... which was the original business problem...

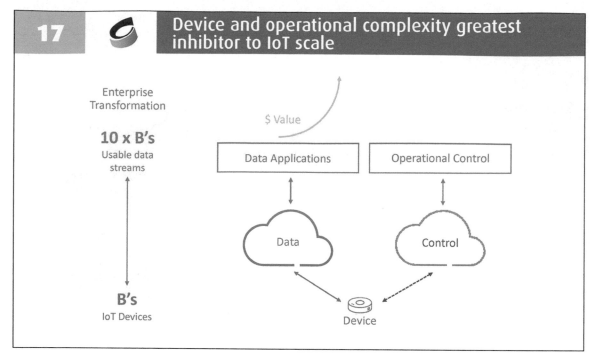

All this is good, but there are some fundamental barriers to realizing IoT and it's all about device and operational complexity.

18 — Scale: Could there be a trillion connected things?

With continuing growth, maybe in 20 years

Key is not number, but scale and exponential growth

1T SENSORS IN 10 YEARS

Year	Unit Price	Units Sold	Industry Revenues	Developed Population	MEMS Rev/ Person	MEMS Unit/ Person
2005	30.000	46,666,667	5,000,000,000	4,000,000,000	1.25	0.01
2010	15.000	466,666,667	7,000,000,000	4,000,000,000	1.75	0.12
2015	1.800	8,333,333,333	15,000,000,000	4,000,000,000	3.75	2.08
2020	0.216	138,888,888,889	30,000,000,000	4,000,000,000	7.50	34.72
2025	0.026	1,388,888,888,889	60,000,000,000	4,000,000,000	15.00	347.22

Chris Wasden at 2014 MEC, via semiwiki

So sensors are the heart of the Internet of Things. And the forecasts for sensing are practically unbelievable. Here's a 1 trillion estimate, and there are many more like that.

The specifics are not as important as the observation of exponential scale and numbers that require automation.

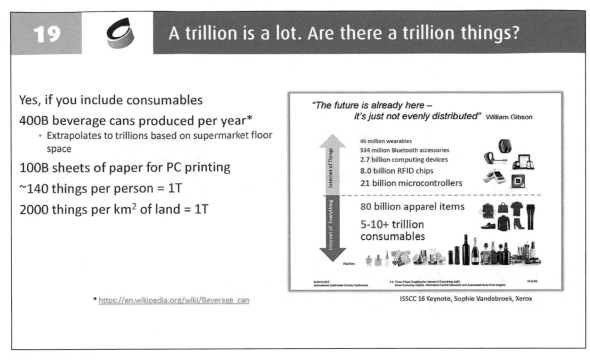

Yes, there are a trillion things, so trillion-scale IoT problems can exist.

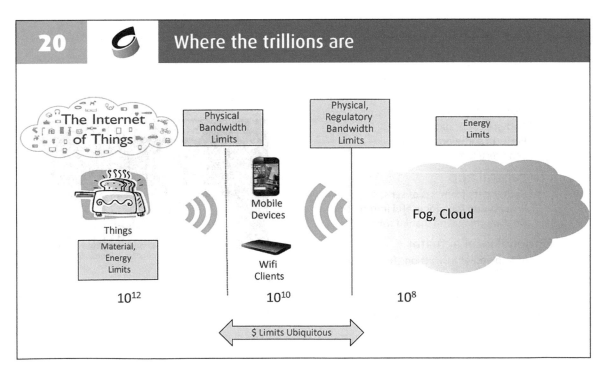

Most of the trillion devices will be at the things at the end of the network. Orders of magnitude fewer as you proceed back toward the cloud. Limits exist everywhere, but especially at the interfaces between gateways and both things and cloud.

21 IoT opportunities –what will the trillions do?

Industry
Commercial Facilities, Warehouses

Home
Smart meters, monitoring

Disaster Management
Landslides, flooding, earthquakes

Public Infrastructure
Management and Maintenance

Automotive
logistics and optimization

Agriculture
Water quality, humiture, soil

Here are some examples of applications that can use huge numbers of connected things.

22 The Basic IOT Algorithm

1. **For each local sensor**
   1. Gather a lot of data
   2. **Perform some processing**
   3. Send results to central app

2. **For central app**
   1. Gather data from lots of sensors
   2. **Process and look for useful information**
   3. Do something with that information

3. **Optional local actuator**
   1. Implement instructions from cloud app

4. **Optional update**
   1. **Update central and/or local apps**

   Red = opportunities for machine learning

Source: imgflip.com

While applications differ, at a high level many can be represented as variants on this theme.

23   Why is ML Moving to the Edge?

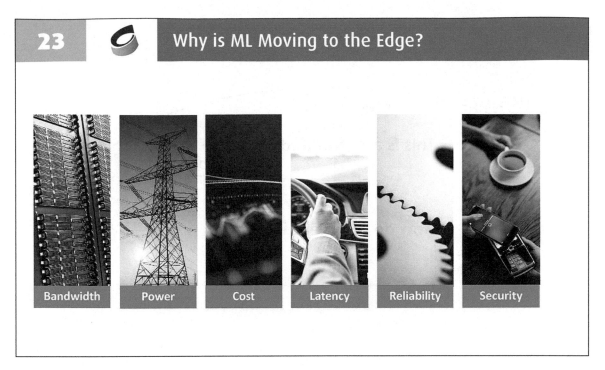

A dvances in compute processing power and AI algorithms have pushed applications, training, and inference to edge devices. It is being driven overall by the laws of physics, laws of economics and laws of the land.

24   Every edge node is constrained

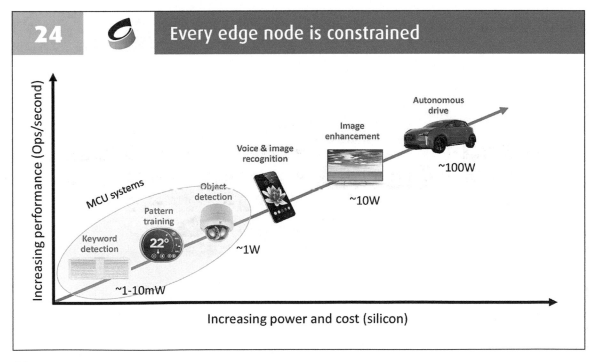

" E dge" is an overloaded term, but regardless of which edge you consider, there are power and cost constraints together with performance requirements.

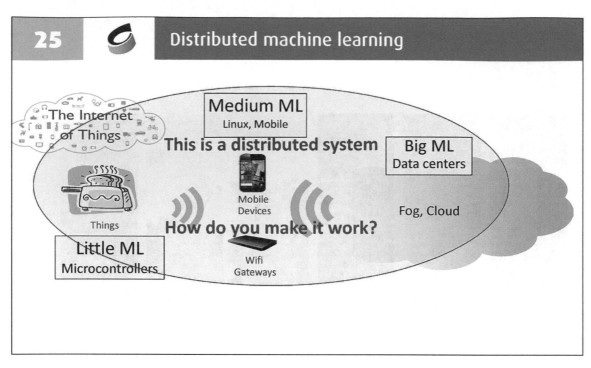

The distributed system is more important than its constituent elements.

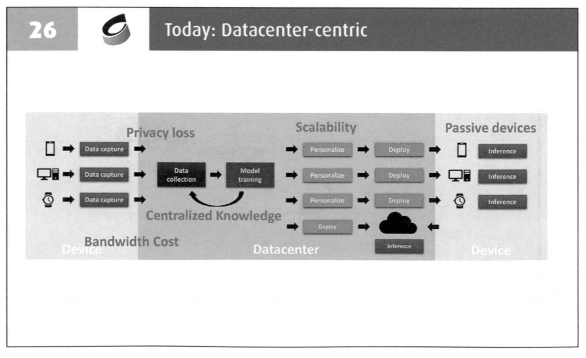

Most of today's systems are oriented around a data center.

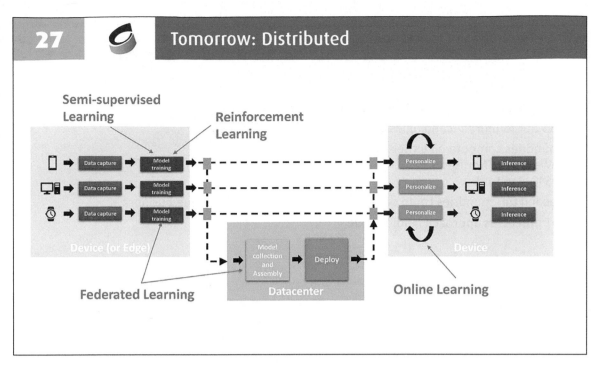

The trend is to a more distributed compute infrastructure.

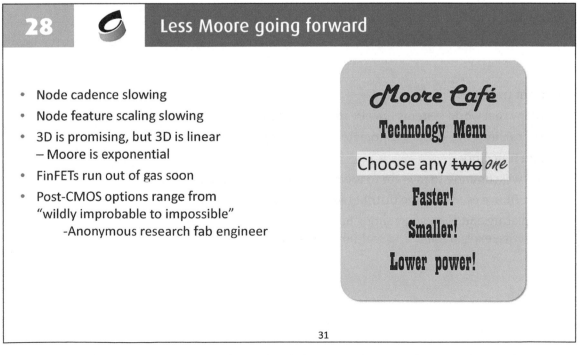

We also need to look at where technology is headed. In particular, Moore's law is slowing down so we can't just hope that future technology will solve are performance and power problems.

29 Thought experiment: What happens when Moore stops?

- "Wright's law" – costs decline as cumulative units increase
- Top tier foundry technologies become essentially equivalent
- Foundries compete on either cost or service
  - Walmart vs Nordstrom
  - $482B vs $14B
- Second tier foundries could catch up, quit, or go into the boutique business
- IoT processes approach leading edge

32

The semiconductor industry (and the broader electronics industry) are built around Moore's law. This will change moving forward. Foundries in particular want to avoid a commodity business, although other sectors show that commodity is where the revenue is.

30 ML and IoT

Current trajectory
- Many real world systems are already automated
- ML can improve existing automated control loops

Future trajectory
- IoT will continue to migrate to today's leading edge processes
- The future of IoT is large distributed systems of constrained devices
- Replacing and even augmenting humans in safety critical control loops requires security, explainability and real-time behavior

How do the IoT and ML trends evolve going forward?

31 Challenge 1: Real time at scale

Amdahl's law still exists – bottlenecks everywhere

Classic methods for speedup

- Removing latency and nondeterminism
- Better coding, compilers, libraries
- Control of interrupts

ML specific additions

- Constrained nets, better training
- Reinforcement learning to shorten feedback loop

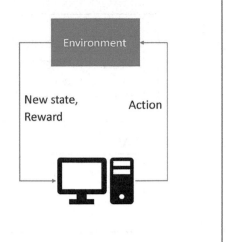

There are several challenges. The first is that most of the world's things have applications where real time behavior matters.

32 Challenge 2: Security (and safety and resilience)

IoT needs to be automated

- A highly automated IoT system needs to be designed to be secure, designed to be safe, and designed to be resilient

Nothing is perfectly secure - need metrics and standards

- Application stack probably not securable
- Target communication & node foundation

Weakest link is a moving target

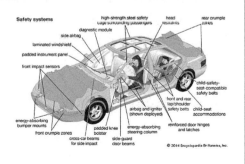

The second challenge is the confluence of sometimes competing requirements for safety, security, resilience, fault tolerance and so on. The history of the auto industry shows us that real solutions will not happen until safety and security are designed in. "Crashing" is a legitimate function of automobiles and all current cars are designed with crashing in mind.

33 — Challenge 3: Explainable AI

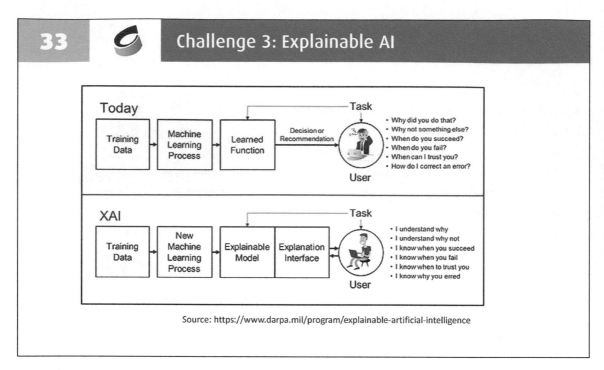

Source: https://www.darpa.mil/program/explainable-artificial-intelligence

It won't be enough to say "the algorithm told us this was the answer" when lives are dependent on the algorithm. Explainable AI is harder than non-explainable, but it's a precursor to widespread deployment.

34 — Properties of distributed, real-time ML systems

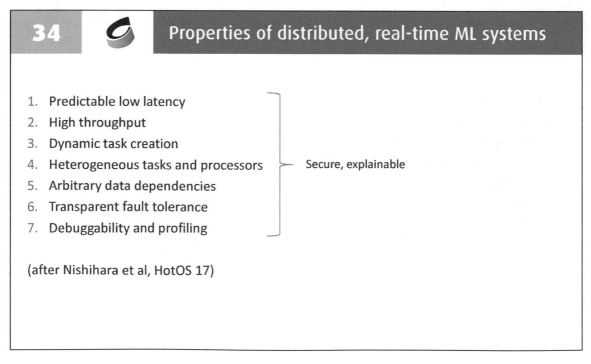

1. Predictable low latency
2. High throughput
3. Dynamic task creation
4. Heterogeneous tasks and processors
5. Arbitrary data dependencies
6. Transparent fault tolerance
7. Debuggability and profiling

Secure, explainable

(after Nishihara et al, HotOS 17)

Nishihara et al. produced a good list of the required properties for distributed real time systems that summarize the requirements of any solution that meets the challenges I just described.

35 Big ML versus Little ML

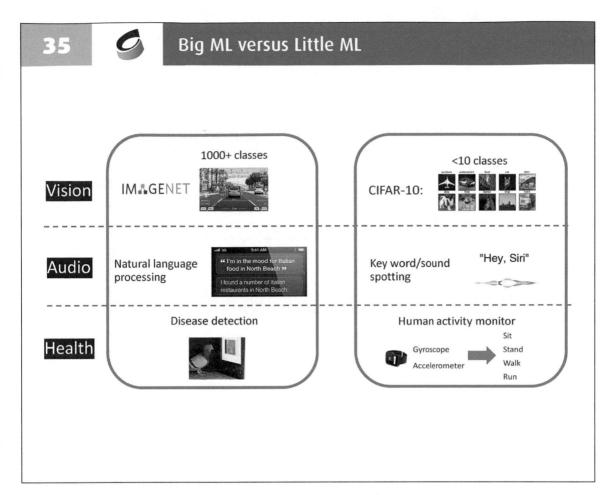

Machine learning has shown superior performance in many different applications. For most applications, there are always relatively easy and hard problems. For vision, Imagenet is probably the most famous example of a hard problem, with more than 1000 image classification classes, including specific types of dogs. There is an easier "mini" version called CIFAR-10, which has only 10 classes with a 32x32 pixel image. The majority of the focus on big-ML is to explore impressive machine learning capabilities which leave people saying "oh, I didn't realize that ML can do this". However, little ML is increasingly important.

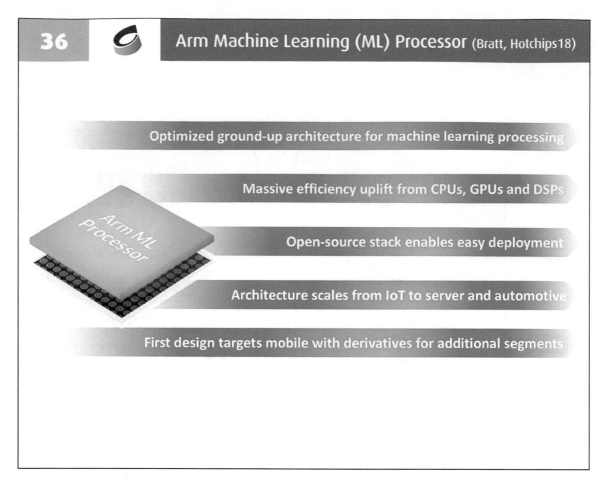

As an example of hardware requirements, we can look at Ian Bratt's presentation from Hotchips 18.

ARM has been closely following the characteristics of ML and AI workloads and how they have changed and proliferated over time.

We looked across all of our IP, and realized that to execute ML workloads most efficiently, we would need to start from scratch, and build a design from the ground-up focused on ML.

That's not to say that we didn't leverage our CPU and GPU expertise. When designing the ML processor, we harnessed our expertise to create a design which is both programmable and high throughput.

Arm's First generation ML processor targets the mobile market. We are targeting the high-end smartphone market because that's where we see innovations taking place first. Then these innovations will trickle down to midrange phones and other consumer devices.

37 Arm's ML Processor

- 16 Compute Engines
- ~ 4 TOP/s of convolution throughput (at 1 GHz)
- Targeting > 3 TOP/W in 7nm and ~2.5mm²
- 8-bit quantized integer support
- 1MB of SRAM
- Support for Android NNAPI and ARMNN
- Optimized for CNNs, RNN support
- To be released 2018

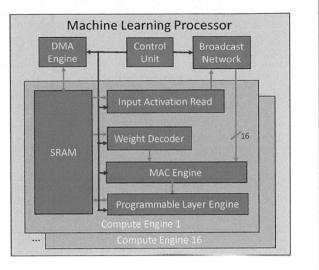

Quantized types work with a defined zero point (defined in 8 bit integer) and a scaling factor for mapping – calculations offset then MAC into i16 or i32 – this allows full hardware support for common quantization schemes.

Hardware requantization is performed after dealing with range scaling.

38 Four Key Ingredients for a Machine Learning Processor (I)

- Static scheduling

- Efficient convolutions

- Bandwidth reduction mechanisms

- Programmability/flexibility

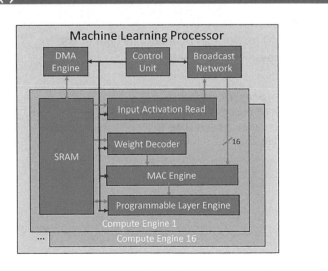

Programmability is key. Data movement and compression for key convolutions is straight-forward.

39 Four Key Ingredients for a Machine Learning Processor (II)

- Static scheduling
- Efficient convolutions
- **Bandwidth reduction mechanisms**
- Programmability/flexibility

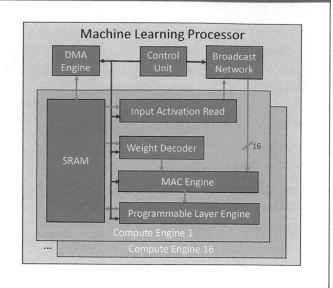

Let's look specifically at memory system bandwidth reduction.

40 Importance of Weight and Feature Map Compression

- **DRAM power can be nearly as high as the processor power itself**
- **ML processor supports**
  - Weight compression
  - Activation compression
  - Tiling

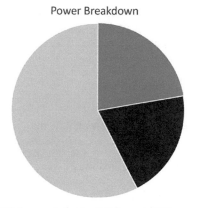

Bandwidth and complexity leads to power.
DDR power can be close to 50% of overall system power.
Compression helps w/ bandwidth.

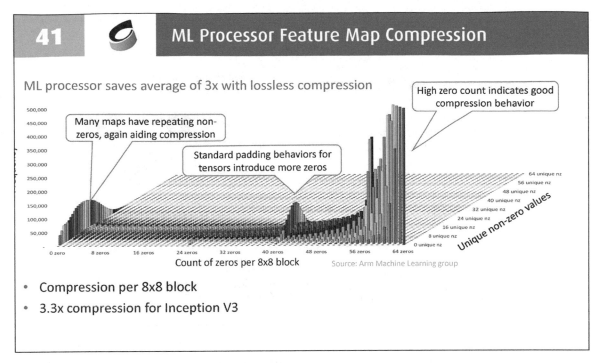

H ere are several places where our data showed that feature map compression could help.

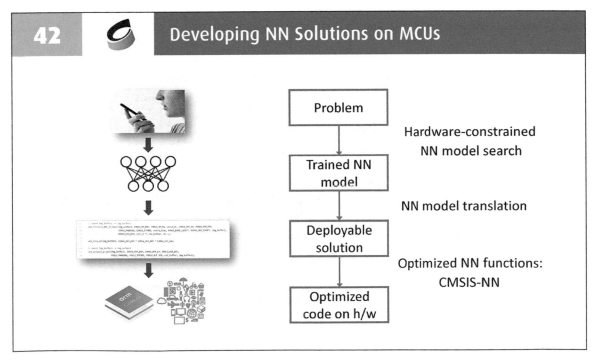

L et's walk through the flow of developing an ML solution for Cortex-M microcontroller systems and explain why/how it is different from other systems.

Let's start with hardware-constrained neural network model search for a "Keyword spotting" use-case.

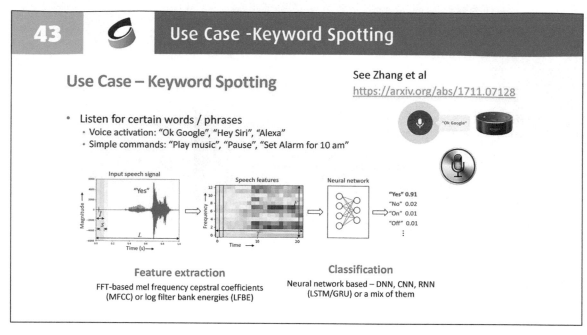

Current deployments of keyword spotting smartphones include: Siri, Google Assistant, Cortana. For assistants there is: Amazon Echo, Google Home. Potential deployments include: smart switches, power outlets, ... Industrial IoT sensors include: gunshot detectors, intruder detectors, ...

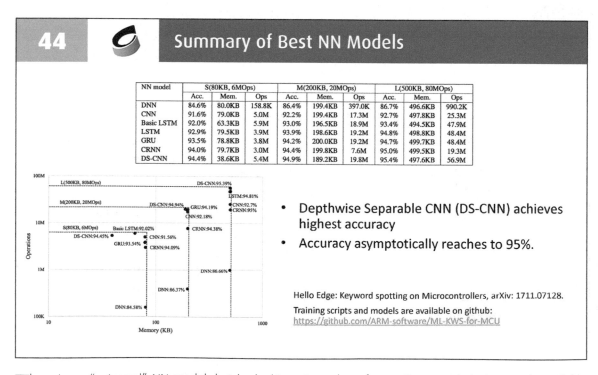

There is no "universal" NN model, but by looking at number of operations needed along with available memory footprint gives a way to find the best one for a given situation.

---

45   Model Deployment on Cortex-M MCUs

- Running ML framework on Cortex-M systems is impractical

- Need to run bare-metal code to efficiently use the limited resources

- Arm NN translates trained model to the code that runs on Cortex-M cores using CMSIS-NN functions

- **CMSIS-NN:** optimized low-level NN functions for Cortex-M CPUs

- CMSIS-NN APIs may also be directly used in the application code

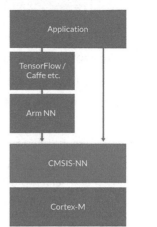

G enerally, models are trained in ML-frameworks and deployed in the same framework environment or a run-time.

ML frameworks or run-time won't fit on Cortex-M systems with very limited memory, so they run bare-metal neural network C or C++ code.

---

46   Convolutional Neural Network (CNN) on Cortex-M7

- CNN with 8-bit weights and 8-bit activations
- Total memory footprint: 87 kB weights + 40 kB activations + 10 kB buffers (I/O etc.)
- Example code available in CMSIS-NN github

NUCLEO-F746ZG
216 MHz, 320 KB SRAM

Layer	Network Parameter	Output activation	Operation count	Runtime on M7
Conv1	5x5x3x32 (2.3 KB)	32x32x32 (32 KB)	4.9 M	31.4 ms
Pool1	3x3, stride of 2	16x16x32 (8 KB)	73.7 K	1.6 ms
Conv2	5x5x32x32 (25 KB)	16x16x32 (8 KB)	13.1 M	42.8 ms
Pool2	3x3, stride of 2	8x8x32 (2 KB)	18.4 K	0.4 ms
Conv3	5x5x32x64 (50 KB)	8x8x64 (4 KB)	6.6 M	22.6 ms
Pool3	3x3, stride of 2	4x4x64 (1 KB)	9.2 K	0.2 ms
ip1	4x4x64x10 (10 KB)	10	20 K	0.1 ms
Total	87 KB weights	Total: 55 KB Max. footprint: 40 KB	24.7 M Ops	99.1 ms

---

H ere are the per-layer results. We use 8-bit weights and 8-bit activations. Total memory footprint is about 130 kB for data.

Total number of ops is about 24.7 million, which takes about 100 ms on the cortex-m7 board we have.

At the top right we see an example of this running with the Cortex-M7 board with an LCD screen.

Cortex-M7 board details: NUCLEO-F746ZG, 216 MHz, 320 kB SRAM 1 MB flash

This demo was performed with a STM32F46G-DISCO board.

47 So where do we go from here?

Need to think of IoT as distributed heterogenous systems, full of AI potential
Current state of AI is not well-suited for Calvinball-style problems
Many problems have constrained subsets where ML works very well
Path forward in ML for IoT involves real-time, explainability, security

Progress is needed on several fronts: there are fundamental research problems yet to be solved in AI, but there are also plenty of opportunities to improve existing ML solutions with technologies that are known now or look like they can be achieved with reasonable effort from where we are now.

48 THE END

The End. For now.

The New Era
of AI Hardware

Jeff Burns

IBM

Chapter 3 by Jeff Burns illustrates IBM's current efforts and future plans for deep learning acceleration. AI systems are necessary to handle the expected 40 trillion gigabytes of data generated by over 30 billion IoT devices in the near term. AI research consists of data requirements, algorithms for data manipulation, and then computing for prediction from the data. Accelerating AI computing involves software, architecture, and hardware development. Accelerating training and inference are two goals of AI hardware research.

Currently available IBM POWER processors were designed with consideration of these goals and large-scale distributing deep learning systems have been demonstrated. IBM has also put forth specialized hardware to accelerate deep learning. The initial accelerator is a digital design based on a new programmable architecture that takes advantage of the reduced precision processing characteristics of deep learning workloads. As the technology marches forward, IBM envisions approximate computing with digital cores first, followed by mixed digital and analog AI cores, and then full analog cores with optimized material technology. Many other industries and universities concur with such a roadmap.

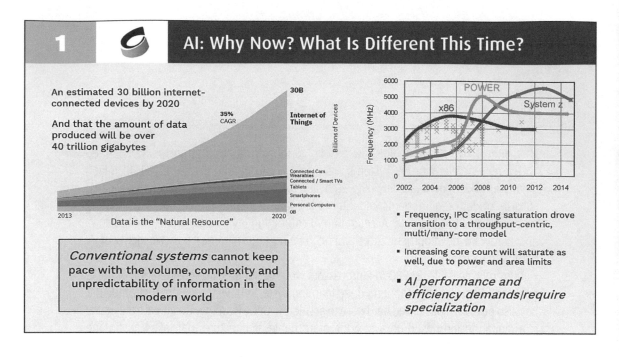

1 **AI: Why Now? What Is Different This Time?**

An estimated 30 billion internet-connected devices by 2020

And that the amount of data produced will be over 40 trillion gigabytes

35% CAGR

30B

Internet of Things

Connected Cars
Wearables
Connected / Smart TVs
Tablets
Smartphones
Personal Computers
0B

2013 — Data is the "Natural Resource" — 2020

Conventional systems cannot keep pace with the volume, complexity and unpredictability of information in the modern world

POWER

x86

System z

Frequency (MHz)

2002 2004 2006 2008 2010 2012 2014

- Frequency, IPC scaling saturation drove transition to a throughput-centric, multi/many-core model

- Increasing core count will saturate as well, due to power and area limits

- *AI performance and efficiency demands require specialization*

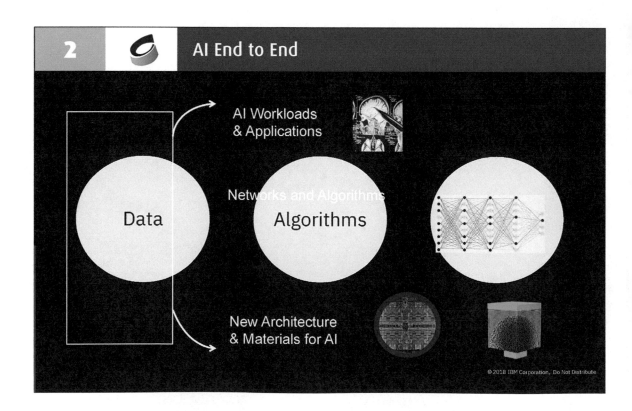

2 **AI End to End**

AI Workloads & Applications

Data

Networks and Algorithms

Algorithms

New Architecture & Materials for AI

© 2018 IBM Corporation. Do Not Distribute

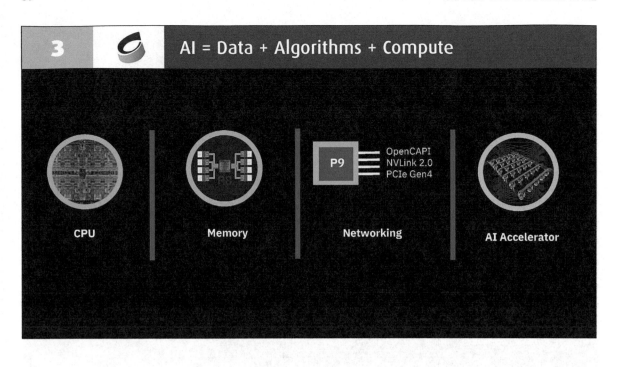

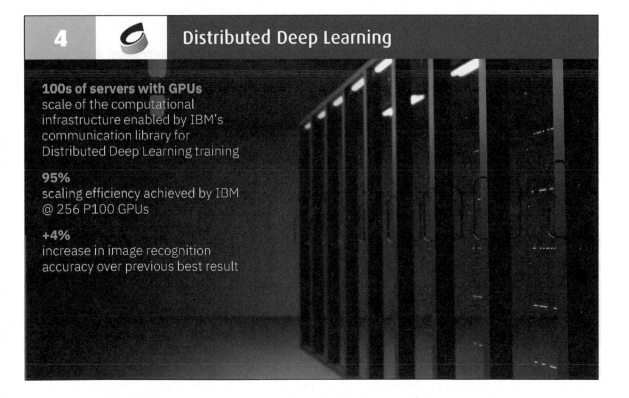

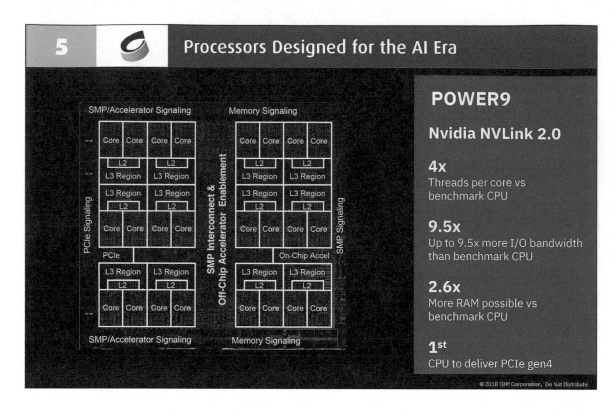

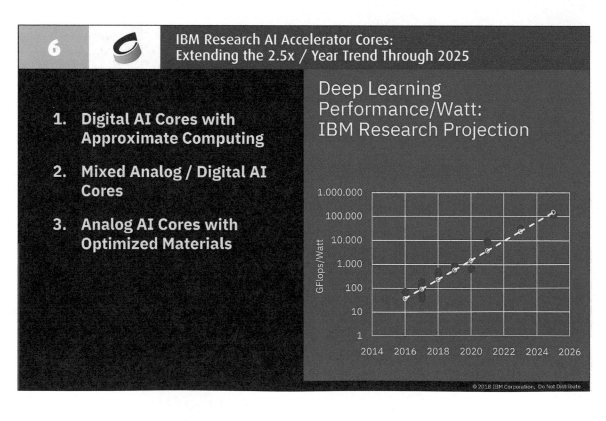

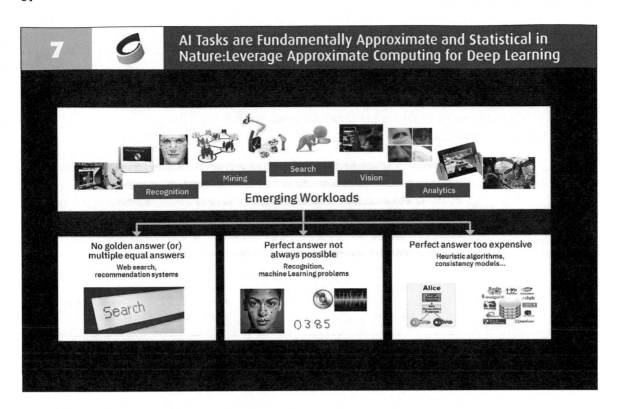

7

AI Tasks are Fundamentally Approximate and Statistical in Nature: Leverage Approximate Computing for Deep Learning

Emerging Workloads

Recognition · Mining · Search · Vision · Analytics

No golden answer (or) multiple equal answers
Web search, recommendation systems

Search

Perfect answer not always possible
Recognition, machine Learning problems

0 3 8 5

Perfect answer too expensive
Heuristic algorithms, consistency models...

Alice

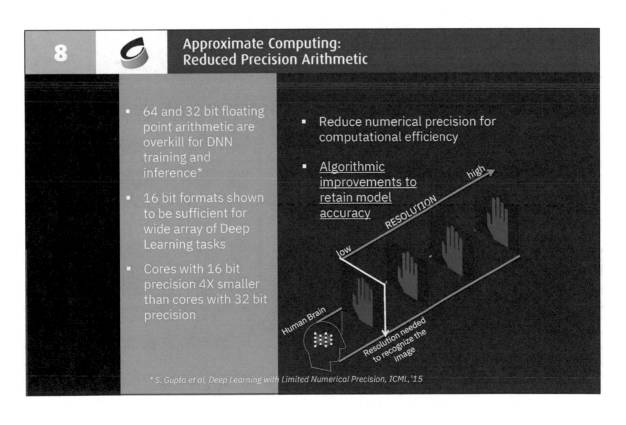

8

Approximate Computing: Reduced Precision Arithmetic

- 64 and 32 bit floating point arithmetic are overkill for DNN training and inference*

- 16 bit formats shown to be sufficient for wide array of Deep Learning tasks

- Cores with 16 bit precision 4X smaller than cores with 32 bit precision

- Reduce numerical precision for computational efficiency

- <u>Algorithmic improvements to retain model accuracy</u>

RESOLUTION — low / high

Human Brain

Resolution needed to recognize the image

*S. Gupta et al, Deep Learning with Limited Numerical Precision, ICML, '15

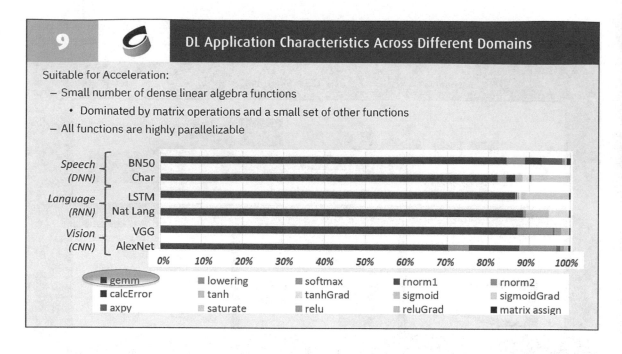

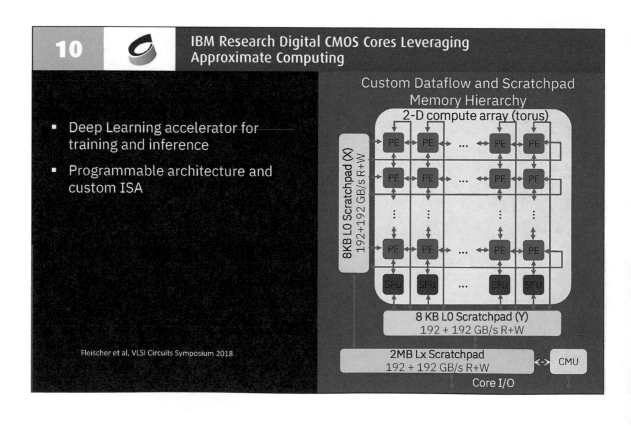

11 IBM Research 14nm 1.5 GHz DL Accelerator Core

- Many highly tuned fp pipelines and high bandwidth throughout **[Performance]**
  - Customized dataflow architectures
  - Algorithm/program/ISA/hardware co-designed specifically for DL
- Balanced multiple-precision support **[Accuracy]**
  - Precision chosen for each computation, for training and inference
- Simplified logic in and around compute pipelines **[Power efficiency]**
  - Carefully curated ISAs, streamlined control logic
- ISA-accessible communications network **[Programmability]**

Peak performance: 1.5 TFLOPS fp16, 12 TOPS ternary, 25 TOPS binary
Sustained utilization: > 90% on multiple neural-network topologies
Core in+out bandwidth: 96+96 GB/s for scalability

"A Scalable Multi-TeraOPS Deep Learning Processor Core for AI Training and Inference", B. Fleischer et al., 2018 Symposia on VLSI Technology and Circuits

"IBM's New Do-It-All AI Chip", Samuel K. Moore, IEEE Spectrum, August 2018, p. 10 - 11

12 Analog AI Cores for In-Memory Computation

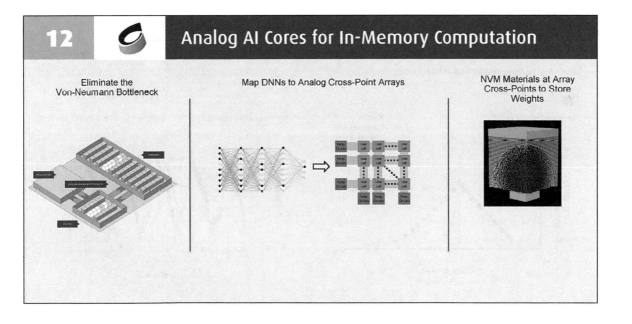

Eliminate the Von-Neumann Bottleneck

Map DNNs to Analog Cross-Point Arrays

NVM Materials at Array Cross-Points to Store Weights

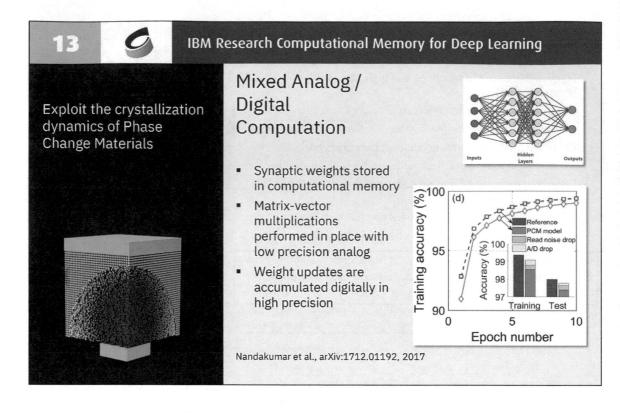

13 IBM Research Computational Memory for Deep Learning

Exploit the crystallization dynamics of Phase Change Materials

Mixed Analog / Digital Computation

- Synaptic weights stored in computational memory
- Matrix-vector multiplications performed in place with low precision analog
- Weight updates are accumulated digitally in high precision

Nandakumar et al., arXiv:1712.01192, 2017

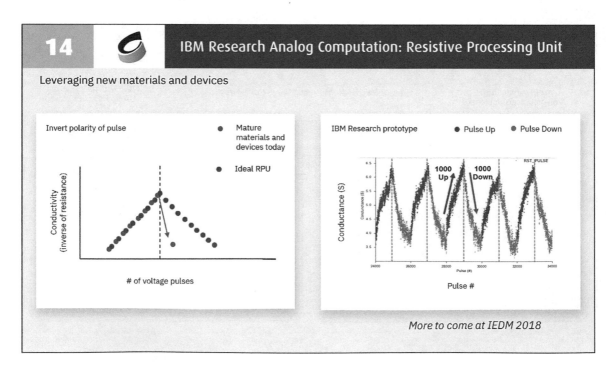

14 IBM Research Analog Computation: Resistive Processing Unit

Leveraging new materials and devices

More to come at IEDM 2018

Orders of magnitude improvements in speed and efficiency are possible

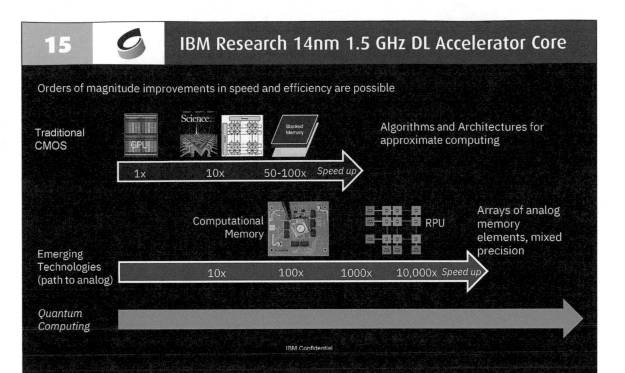

Traditional
CMOS

Algorithms and Architectures for
approximate computing

1x   10x   50-100x   *Speed up*

Computational
Memory

RPU

Arrays of analog
memory
elements, mixed
precision

Emerging
Technologies
(path to analog)

10x   100x   1000x   10,000x *Speed up*

*Quantum
Computing*

IBM Confidential

CHAPTER 04

AI and the Opportunity for Unconventional Computing Platforms

Naveen Verma

Princeton University

This chapter by Naveen Verma discusses a case for circuit and architectural approaches for in-memory computing, which is another topic of high interest to the community. This approach opens up the opportunity of bottoms-up hardware design with the potential for significant acceleration. Due to the high cost of data movement, the potential gain from in-memory computing is projected to be 10x in performance and 10x in efficiency. Similar to a prior chapter, this chapter emphasizes the impact of approximate computing using different input bit selection. The training error can be reduced using optimal statistical regression techniques and appropriate bit size.

Verma shows how conventional memories can be used for in-memory computation, e.g., SRAM. As an example, it is shown that a 6T-cell SRAM array can be used as a weak classifier, as demonstrated on MNIST. Using capacitors, a high-density bit cell can also be realized. Measurement results are presented from several fabricated chips providing compelling evidence for in-memory computing potential. This work also illustrates how university research plays a crucial part in realizing the industry vision.

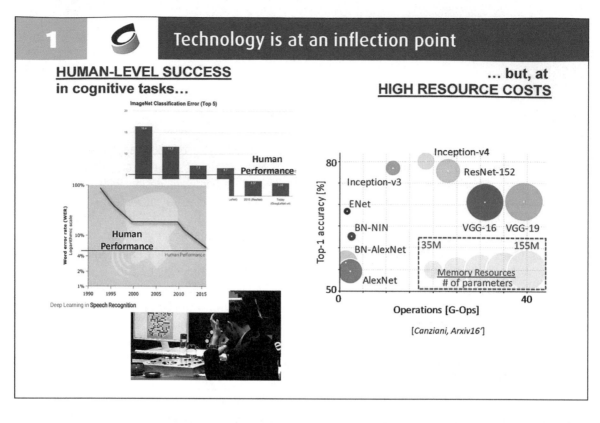

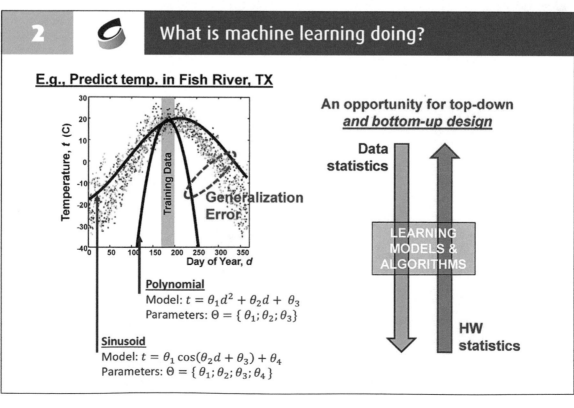

3   E.g.: reducing compute precision

Linear classifier with 4b weights:

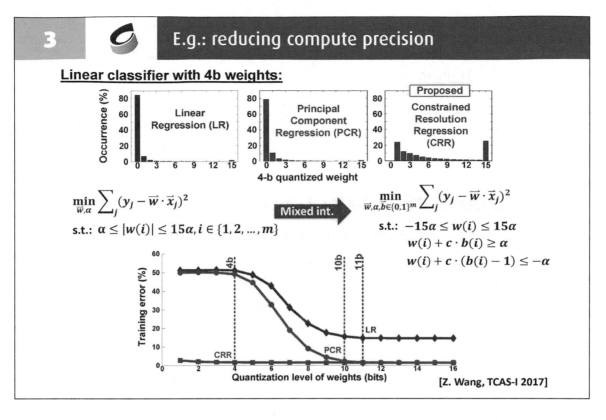

$$\min_{\vec{w},\alpha} \sum_j (y_j - \vec{w} \cdot \vec{x}_j)^2$$

s.t.: $\alpha \le |w(i)| \le 15\alpha, i \in \{1, 2, \dots, m\}$

Mixed int.

$$\min_{\vec{w},\alpha,\vec{b} \in \{0,1\}^m} \sum_j (y_j - \vec{w} \cdot \vec{x}_j)^2$$

s.t.: $-15\alpha \le w(i) \le 15\alpha$

$w(i) + c \cdot b(i) \ge \alpha$

$w(i) + c \cdot (b(i) - 1) \le -\alpha$

[Z. Wang, TCAS-I 2017]

4   E.g.: picking easier computations

Error-Adaptive Classifier Boosting:

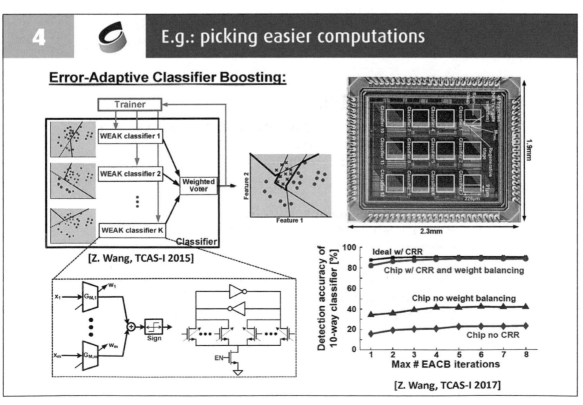

[Z. Wang, TCAS-I 2015]

[Z. Wang, TCAS-I 2017]

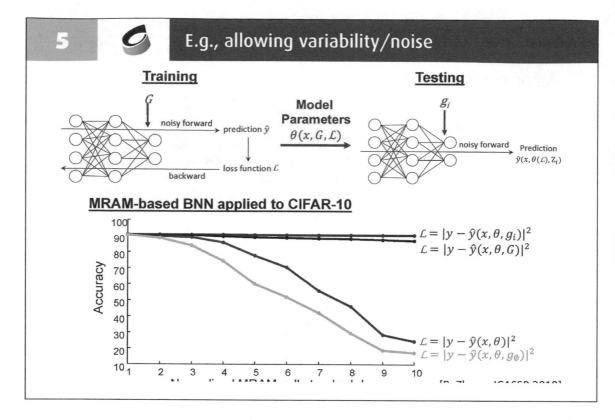

5   E.g., allowing variability/noise

Training   **Testing**

Model Parameters
$\theta(x, G, \mathcal{L})$

noisy forward → prediction $\hat{y}$

loss function $\mathcal{L}$

backward

noisy forward → Prediction $\hat{y}(x, \theta(\mathcal{L}), z_i)$

MRAM-based BNN applied to CIFAR-10

$\mathcal{L} = |y - \hat{y}(x, \theta, g_i)|^2$
$\mathcal{L} = |y - \hat{y}(x, \theta, G)|^2$

$\mathcal{L} = |y - \hat{y}(x, \theta)|^2$
$\mathcal{L} = |y - \hat{y}(x, \theta, g_\phi)|^2$

---

6   Architecture design

System Pain Points:

1. **Inst. & data conversion** [J. Zhang, ISSCC'15], [Z. Wang, TCAS-I'17]
   - E/MAC (12b, 45nm): ~1pJ
   - E/ADC (12b, 45nm): **~200pJ**

2. **Enabling heterogeneity/specialization** [H. Jia, VLSI'17]
   - Programmability **opposes specialization**
   - Programming accelerators is very difficult

3. **Memory accessing** [J. Zhang, VLSI'16][H. Valavi, VLSI'18]
   - E/SRAM (10/32/1000 kB, 45nm): **~10/20/100 pJ**

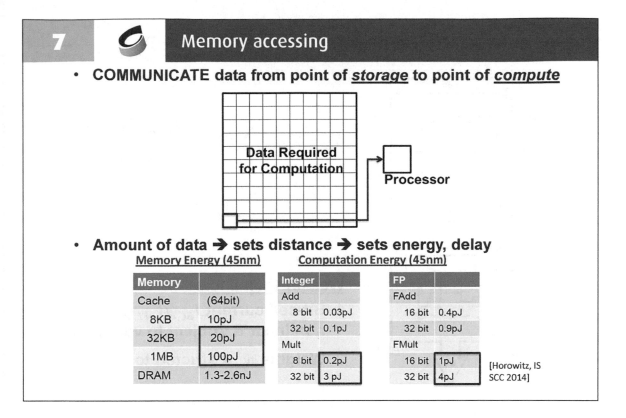

7 Memory accessing

- **COMMUNICATE data from point of _storage_ to point of _compute_**
- **Amount of data → sets distance → sets energy, delay**

Memory Energy (45nm)

Memory	
Cache	(64bit)
8KB	10pJ
32KB	20pJ
1MB	100pJ
DRAM	1.3-2.6nJ

Computation Energy (45nm)

Integer		FP	
Add		FAdd	
8 bit	0.03pJ	16 bit	0.4pJ
32 bit	0.1pJ	32 bit	0.9pJ
Mult		FMult	
8 bit	0.2pJ	16 bit	1pJ
32 bit	3 pJ	32 bit	4pJ

[Horowitz, ISSCC 2014]

8 Classifier in standard 6T SRAM

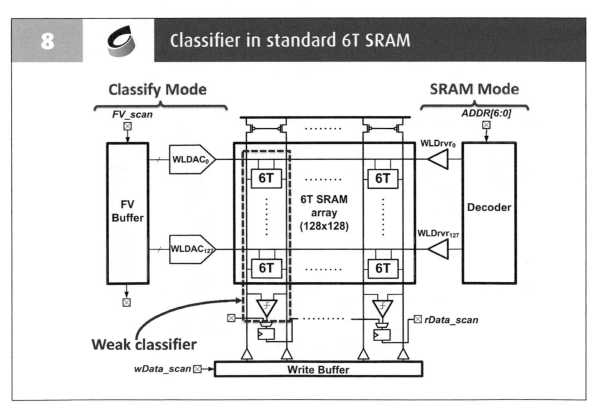

9 Current-domain in-memory computing

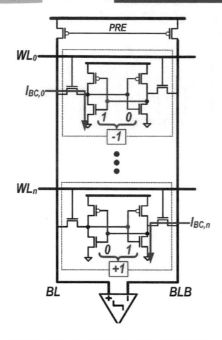

$$V_{BL} - V_{BLB} = \sum_{n=0}^{127} w_n \times \frac{\int I_{BC,n}(V_{WL,n})\,dt}{C_{BL/BLB}}$$

$$= \sum_{n=0}^{127} w_n \times \Delta V_{BL/BLB,n}$$

Weaker than linear:

1. Bit-cell current $I_{BC,n}(V_{WL,n})$ nonlinear & subject to variation

2. Weights w_n restricted to +/-1

10 Training overview

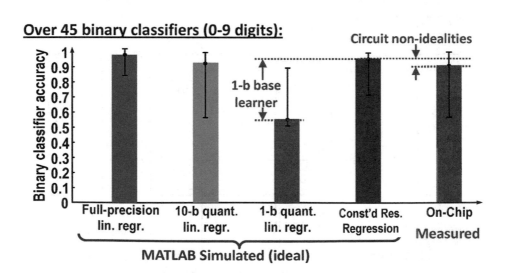

Over 45 binary classifiers (0-9 digits):

Binary classifier accuracy

Circuit non-idealities

1-b base learner

- Full-precision lin. regr.
- 10-b quant. lin. regr.
- 1-b quant. lin. regr.
- Const'd Res. Regression
- On-Chip Measured

MATLAB Simulated (ideal)

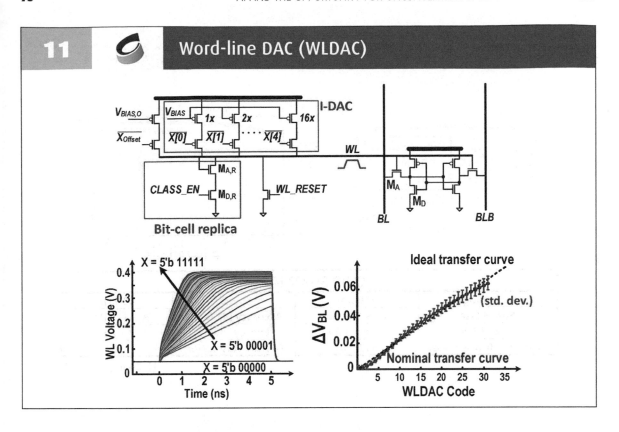

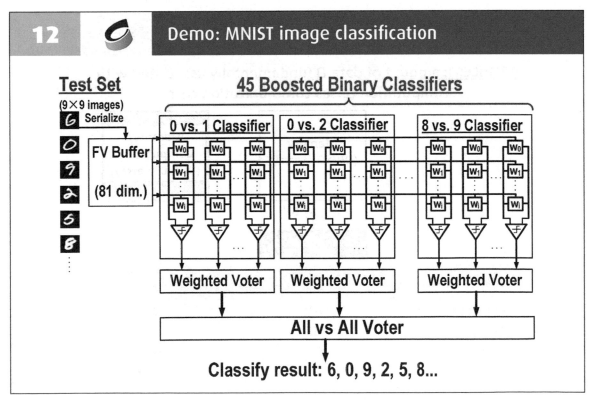

13 Prototype

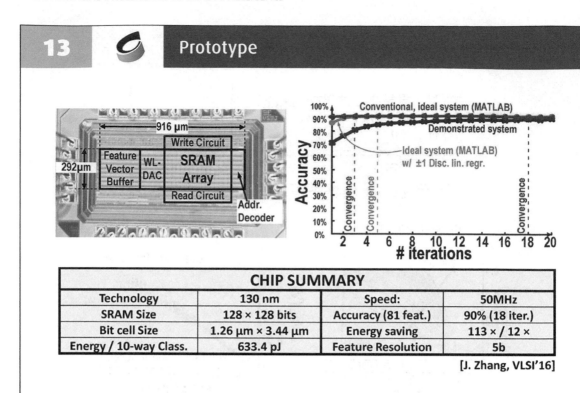

CHIP SUMMARY			
Technology	130 nm	Speed:	50MHz
SRAM Size	128 × 128 bits	Accuracy (81 feat.)	90% (18 iter.)
Bit cell Size	1.26 μm × 3.44 μm	Energy saving	113 × / 12 ×
Energy / 10-way Class.	633.4 pJ	Feature Resolution	5b

[J. Zhang, VLSI'16]

14 New system tradeoffs --SNR

PROBLEM: amount of data *D* fundamentally associated with computation raises communication cost.

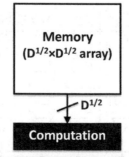

Metric (for entire compute)	Traditional	In-memory
Bandwidth	$1/D^{1/2}$	1
Latency	D	1
Energy	$D^{3/2}$	$\sim D$
SNR	1	$\sim 1/D^{1/2}$

15   Where does in-memory computing stand? (I)

- **Potential for 10× higher performance, 10× higher efficiency…**

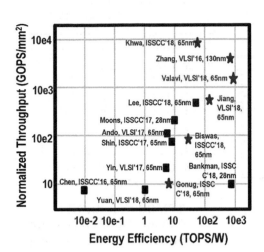

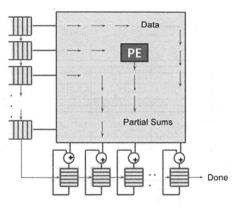

Digital PE
- E_{BUFF} = 250 fJ
- $E_{4b\text{-}MAC}$ = 150 fJ
- E_{COMM} = 40 fJ

} **Bit-cell PE** 30 fJ

16   Where does in-memory computing stand? (II)

- **Scale?**

- **Programmability?**

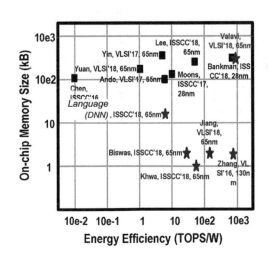

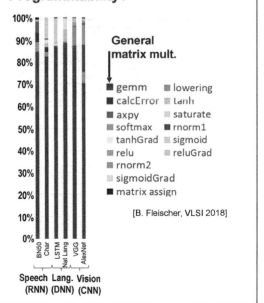

General matrix mult.

- gemm
- calcError
- axpy
- softmax
- tanhGrad
- relu
- rnorm2
- sigmoidGrad
- matrix assign
- lowering
- tanh
- saturate
- rnorm1
- sigmoid
- reluGrad

[B. Fleischer, VLSI 2018]

17 · Charge-domain in-memory computing

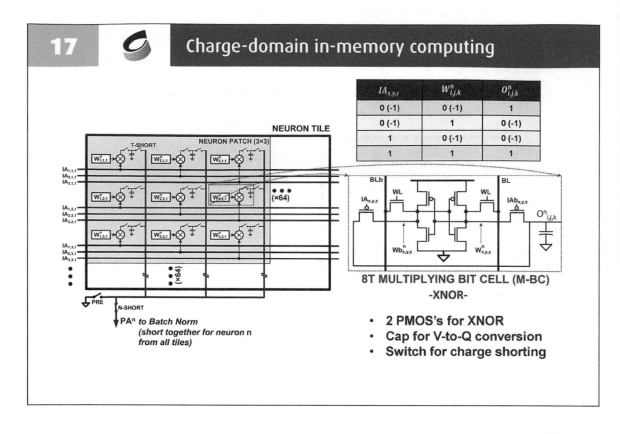

$IA_{x,y,z}$	$W^n_{i,j,k}$	$O^n_{i,j,k}$
0 (-1)	0 (-1)	1
0 (-1)	1	0 (-1)
1	0 (-1)	0 (-1)
1	1	1

8T MULTIPLYING BIT CELL (M-BC)
-XNOR-

- **2 PMOS's for XNOR**
- **Cap for V-to-Q conversion**
- **Switch for charge shorting**

18 · High-density bit cells

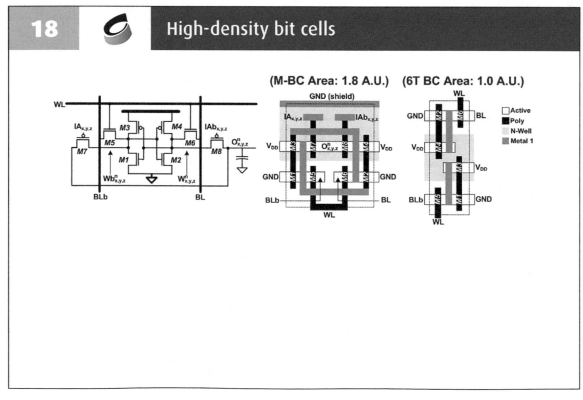

(M-BC Area: 1.8 A.U.) **(6T BC Area: 1.0 A.U.)**

19 — Highly process/temp. stable accumulate

E.g., MOM-capacitor matching (130nm):

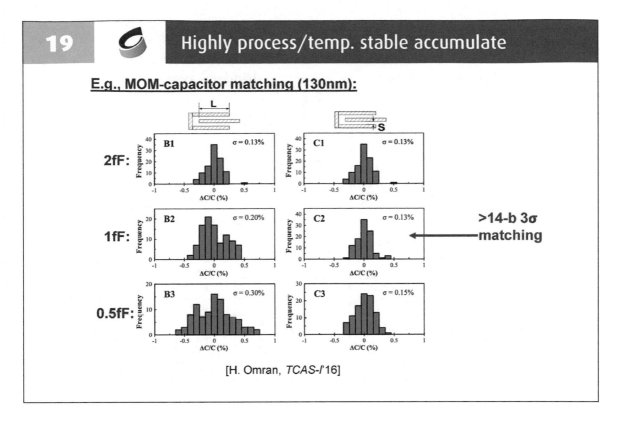

[H. Omran, *TCAS-I'*16]

20 — Pipeline architecture for CNNs

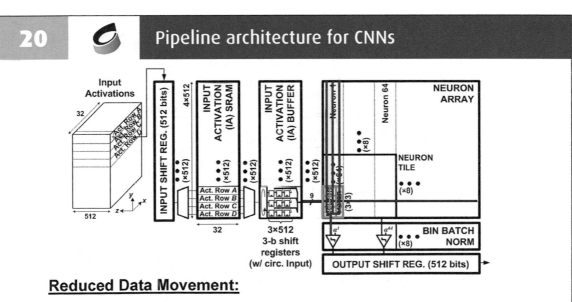

Reduced Data Movement:

1. **Input activations broadcast minimal distance over dense M-BCs**

2. Weights are stationary in M-BCs

3. **High dynamic range analog pre-activation computed via single-bit control**

21   64-bank, 2.4Mb array prototype

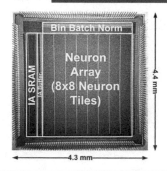

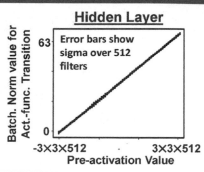

Hidden Layer

Error bars show sigma over 512 filters

Batch. Norm value for Act.-func. Transition

Pre-activation Value: $-3\times3\times512$ to $3\times3\times512$

	Chen, ISSCC'16	Moons, ISSCC'17	Bang, ISSCC'17	Ando, VLSI'17	Bankman, ISSCC'18	This Work
Technology	65nm	28nm	40nm	65nm	28nm	65nm
Area (mm²)	16	1.87	7.1	12	6	17.6
Operating VDD	0.8-1.2	1	0.63-0.9	0.55-1	0.8 (0.6)	0.94/0.68/1.2
Bit precision	16b	4-16b	6-32b	1b	1b	1b
on-chip Mem.	108kB	128kB	270kB	100kB	328kB	295kB
Throughput (GOPS)	120	400	108	1264	400 (60)	18876
TOPS/W	0.0096	10	0.384	6	532 (772)	866

22   Neural-network demonstrations

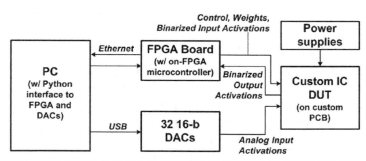

	MNIST	CIFAR-10	SVHN
Test Acc. (Chip/SW)	98.60/98.92	83.27/83.50	94.35/95.10
Validation Acc. (Chip/SW)	98.58/98.75	84.09/84.37	94.03/94.63
Baseline Neural Network	64 CONV3 – BN 64 CONV3 – BN 128 CONV3 – BN 128 CONV3 – BN 10 FC	64 CONV3 – BN 64 CONV3 – BN 128 CONV3 – BN 128 CONV3 – BN 256 CONV3 – BN 256 CONV3 – BN 1024 FC 1024 FC 10 FC	64 CONV3 – BN 64 CONV3 – BN 128 CONV3 – BN 128 CONV3 – BN 256 CONV3 – BN 256 CONV3 – BN 1024 FC 1024 FC 10 FC
E/Class. (CONV. layers)	0.8μJ	3.55μJ	3.55μJ

23   Toward programmable in-memory computing

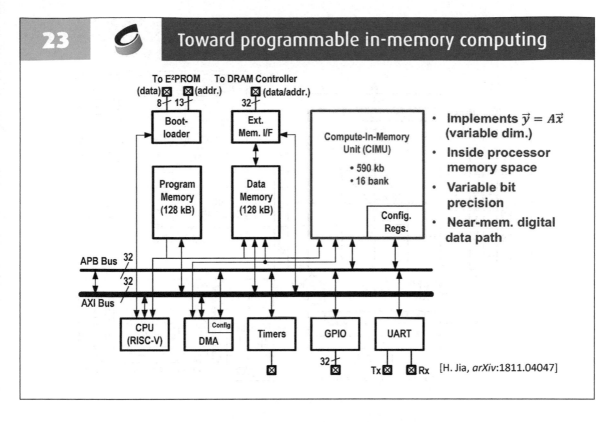

[H. Jia, *arXiv*:1811.04047]

24   Bit-parallel/bit-serial compute

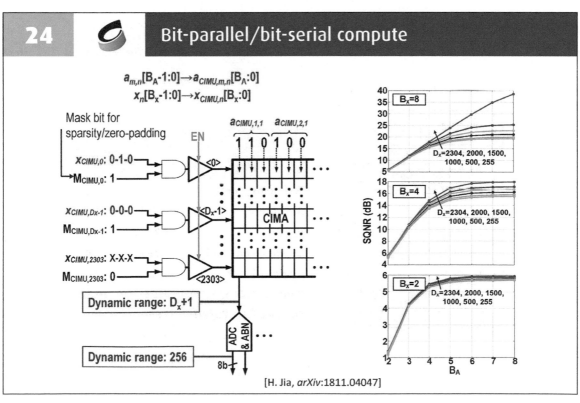

[H. Jia, *arXiv*:1811.04047]

25 — Integrated near-memory data path

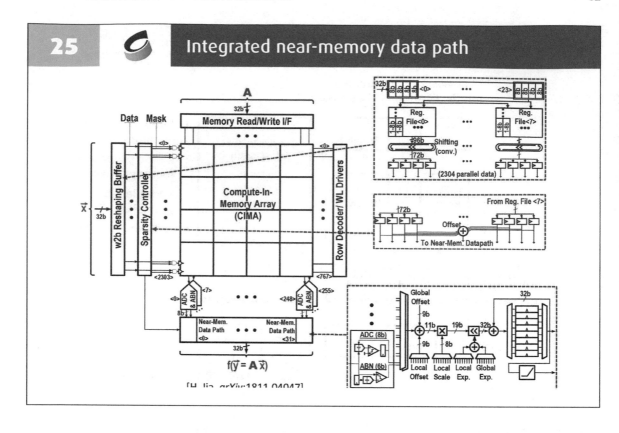

[H. Jia, arXiv:1811.04047]

26 — Programmable in-memory computing

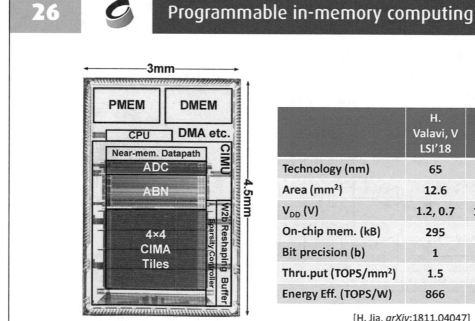

	H. Valavi, V LSI'18	This work
Technology (nm)	65	65
Area (mm²)	12.6	8.5
V_{DD} (V)	1.2, 0.7	1.2, 0.85
On-chip mem. (kB)	295	74
Bit precision (b)	1	1-8
Thru.put (TOPS/mm²)	1.5	0.22
Energy Eff. (TOPS/W)	866	300

[H. Jia, arXiv:1811.04047]

27   Measured transfer functions

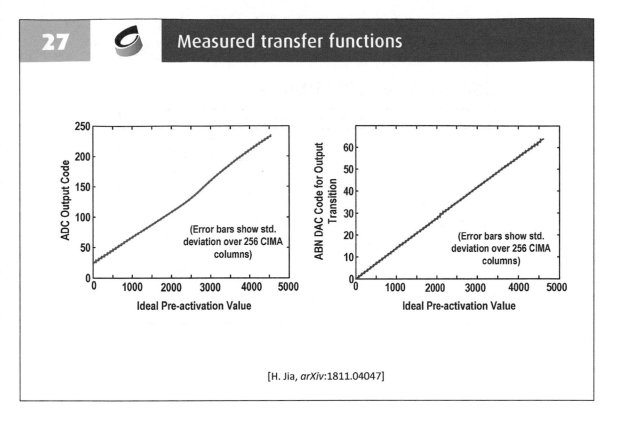

[H. Jia, *arXiv*:1811.04047]

28   Measured SQNR

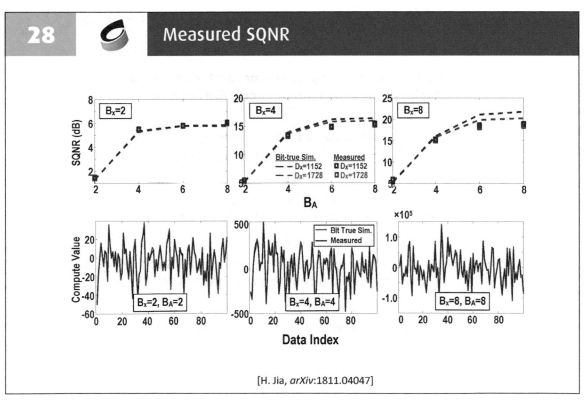

[H. Jia, *arXiv*:1811.04047]

29 Neural-network demonstrations

CNNs applied to CIFAR-10 Dataset:

Neural-Network Demonstrations		
	Network A (4/4-b activations/weights)	**Network B** (1/1-b activations/weights)
Accuracy of chip (vs. SW sim'ed)	92.4% (vs. 92.7%)	89.3% (vs. 89.8%)
Energy/10-way Class.[1]	105.2 µJ	5.31 µJ
Throughput[1]	23 images/sec.	176 images/sec.
Neural Network Topology	L1: 128 CONV3 – Batch norm L2: 128 CONV3 – POOL – Batch norm. L3: 256 CONV3 – Batch. norm L4: 256 CONV3 – POOL – Batch norm. L5: 256 CONV3 – Batch norm. L6: 256 CONV3 – POOL – Batch norm. L7-8: 1024 FC – Batch norm. L9: 10 FC – Batch norm.	L1: 128 CONV3 – Batch Norm. L2: 128 CONV3 – POOL – Batch Norm. L3: 256 CONV3 – Batch Norm. L4: 256 CONV3 – POOL – Batch Norm. L5: 256 CONV3 – Batch Norm. L6: 256 CONV3 – POOL – Batch Norm. L7-8: 1024 FC – Batch norm. L9: 10 FC – Batch norm.

[H. Jia, *arXiv*:1811.04047]

30 Summary & conclusions

AI systems are pushing traditional architectures and technologies to limits of energy/throughput

Statistical learning opens up the opportunity for <u>bottom-up</u> design

Unconventional architectures and technologies can substantially overcome the bottlenecks in today's systems

How to exploit unconventional architectures and technologies must be considered on the system level

The entire designer/user stack must be supported

Acknowledgments: STARnet (SONIC, C-FAR), JUMP (C-BRIC), DARPA (FRANC), ADI.

CHAPTER 05

Thermodynamic Computing

Todd Hylton

UC San Diego

This chapter by Todd Hylton provides an out-of-the-box proposal for thermodynamic computing. This new computing paradigm is based on thermodynamically evolving computing systems that can be biased through programming, training, and rewarding. The premise behind thermodynamic computing is that striving for thermodynamic efficiency is not only highly desirable in hardware components but may also be used as an embedded capability in the creation of algorithms. Can dissipated heat be used to trigger adaptation/restructuring of (parts of) the functioning hardware, thus allowing hardware to evolve increasingly efficient computing strategies? Hylton puts forth non-equilibrium thermodynamic concepts that drive the organization of open systems as a natural response to external input potentials.

The architecture of a thermodynamic computer might consist of a generic fabric of thermodynamically evolvable cores embedded in a reconfigurable network of connections. Such systems would adapt as they dissipate energy, enter low dissipation homeostatic states and as a result 'learn' to 'predict' future inputs. If realized, thermodynamic computers may offer much greater efficiency of computation as well as provide new ways to perform algorithmic techniques within areas like machine learning and neuromorphic computing.

MOTIVATION

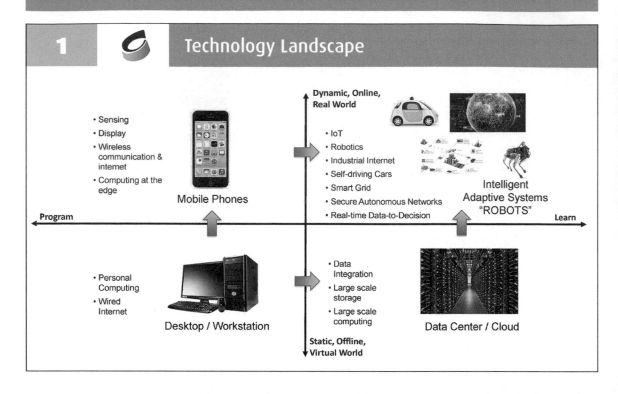

1 — Technology Landscape

- Sensing
- Display
- Wireless communication & internet
- Computing at the edge

Mobile Phones

Dynamic, Online, Real World

- IoT
- Robotics
- Industrial Internet
- Self-driving Cars
- Smart Grid
- Secure Autonomous Networks
- Real-time Data-to-Decision

Intelligent Adaptive Systems "ROBOTS"

Program — Learn

- Personal Computing
- Wired Internet

Desktop / Workstation

- Data Integration
- Large scale storage
- Large scale computing

Data Center / Cloud

Static, Offline, Virtual World

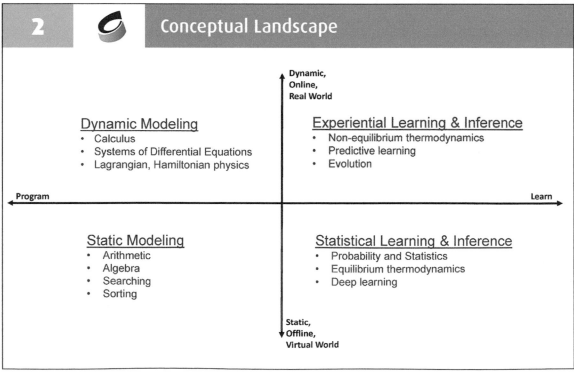

2 — Conceptual Landscape

Dynamic, Online, Real World

Dynamic Modeling
- Calculus
- Systems of Differential Equations
- Lagrangian, Hamiltonian physics

Experiential Learning & Inference
- Non-equilibrium thermodynamics
- Predictive learning
- Evolution

Program — Learn

Static Modeling
- Arithmetic
- Algebra
- Searching
- Sorting

Statistical Learning & Inference
- Probability and Statistics
- Equilibrium thermodynamics
- Deep learning

Static, Offline, Virtual World

3 The Problem

- The primary problem in computing today is that computers cannot organize themselves.
  - Trillions of degrees of freedom doing the same stuff over and over...
  - Narrowly focused AI capabilities
  - Using lots of energy

- Our mechanistic approach to the problem is ill-suited to complex, real-world problems.
  - Machines are the sum of their parts
  - Machines are disconnected from the world except through us
  - Machines cannot evolve without us
  - The world is not a machine

4 Thermodynamic Computing Hypothesis

- Why Thermodynamics?
  - Thermodynamics is *universal*
  - Thermodynamics is *temporal*
  - Thermodynamics is *efficient*
  - Thermodynamics is *not mechanistic*
- *Thermodynamics drives the evolution of everything*
- *Thermodynamic evolution* is the missing, unifying concept in computing (and in many other domains)
- Electronic systems are extraordinarily well-suited for thermodynamic evolution

Thermodynamics should be the principal concept in future computing systems

THERMODYNAMICS AND EVOLUTION

5 — Thermodynamics of Open Systems

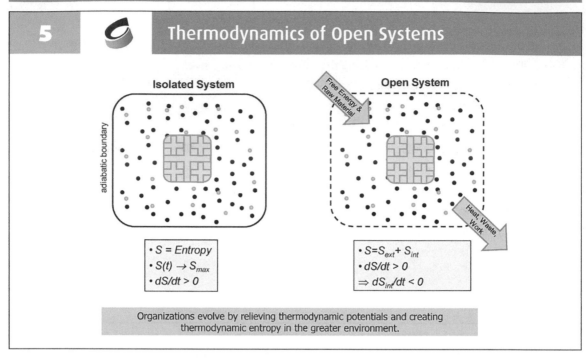

Organizations evolve by relieving thermodynamic potentials and creating thermodynamic entropy in the greater environment.

6 — Thermodynamically Evolved Structures

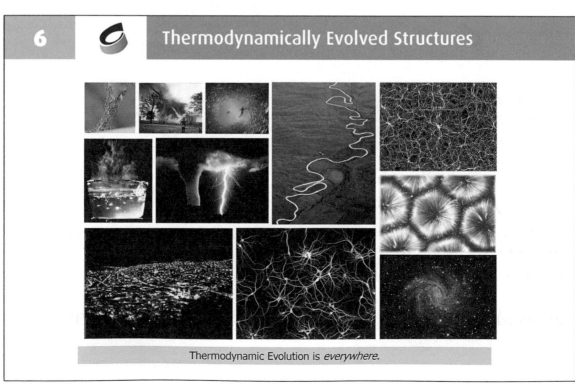

Thermodynamic Evolution is *everywhere*.

7 Example: Iron Filings Near a Magnet

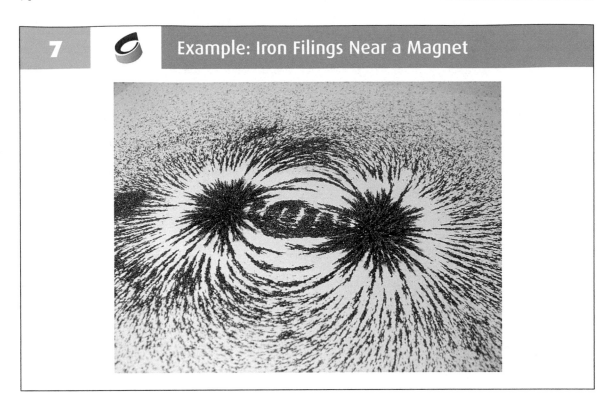

8 ArbortronDemos –Stanford Complexity Group

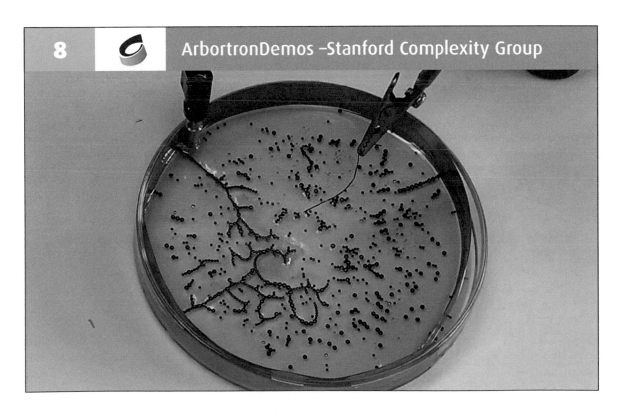

9 Life and Modern Electronic Systems

Requirement	Life	Electronics
Environment	Chemical Potential	Electric Potential
Potential source	Sun / Earth / ...	Power Supplies / Inputs
Room Temperature	Yes	Yes
Interaction energies	~0.1-10 V	~0.1-10 V
Interaction scale	Molecular	Nanometer
Result	thermodynamically evolving biological systems	thermodynamically evolving electronic systems?

The energy scale of excitations in biological and electronic systems are similar because they both derive from the electronic structure of the materials of which they are composed. *Electronic systems do not evolve today because we design them so that they do not evolve.*

10 Thermodynamic Evolution Hypothesis

- Thermodynamic evolution supposes that all organization spontaneously emerges in order to use sources of free energy in the universe.
  - Thermodynamic evolution is second law of thermodynamics, except that it adds the idea that organization spontaneously emerges to access & dissipate free energy.
  - The first law of thermodynamics implies that there is competition for energy and that conservation laws apply all to all evolution.

- Thermodynamic evolution, like Darwinian evolution, is an argument of self-consistency and not cause-and-effect.
  - An organism / organization is "fit because it survives and survives because it is fit".
  - Thermodynamic evolution is not mechanistic – there is no universal "formula"

EQUILIBRIUM THERMODYNAMIC FLUCTUATION DISSIPATION THEORY

11 Non-equilibrium Fluctuation Theorems

- Generalization of the second law of thermodynamics for driven, non-equilibrium systems; useful tool to study self-organized systems.

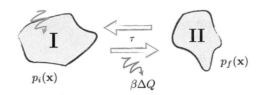

$$\frac{\pi(II^* \to I^*)}{\pi(I \to II)} = \left\langle e^{\ln\left[\frac{p_f(x)}{p_i(x)}\right]} \left\langle e^{-\beta\Delta Q}\right\rangle\right\rangle_{I \to II}$$

- $\pi(I \to II)$ is the probability of the transition from macrostate I to II when the system is driven by external driving signal
- The relative probability of a thermodynamic trajectory compared to its reverse is related to the average energy dissipated into a thermal reservoir.

Perunov, Nikolay, Robert A. Marsland, and Jeremy L. England. "Statistical physics of adaptation." *Physical Review X* 6.2 (2016): 021036.

12 Reliable High Dissipation

- Macrostate II is more likely than III if

$$\Delta Q_{I \to II} >> \Delta Q_{I \to III}$$

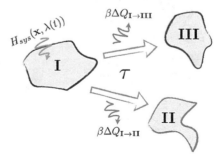

- Dissipation proportional to increased correlation with driving signal

$$\Delta Q \propto \Delta I^{\mathcal{R}_H \mathcal{S}}$$

- Dissipation can enhance correlation – i.e. cause adaptation / learning.

13 Reliable Low Dissipation

- Fluctuation theorem for homeostasis of macrostate I $$\left\langle e^{\ln\left[\frac{p_f(x)}{p_i(x)}\right]}\langle e^{-\beta\Delta Q}\rangle\right\rangle_{I\to I}=1$$

- Reliable low dissipation condition is equivalent to Information Bottleneck

$$Max_{p(k_0|i_H)}\mathcal{I}^{\mathcal{R}_1\mathcal{S}_0}-\lambda\mathcal{I}^{\mathcal{R}_H\mathcal{S}_0}$$

- $\mathcal{I}^{\mathcal{R}_H\mathcal{S}_0}$ Measure of memory/complexity; $\mathcal{I}^{\mathcal{R}_1\mathcal{S}_0}$ Measure of prediction

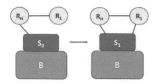

- Solution is $$p(k_0|i_H)\propto e^{\frac{-1}{\lambda}(D_{KL}[p(i_1|i_H)|p(i_1|k_0)])}$$

Energy efficient dynamics implies predictive behavior.

Ganesh, Natesh. "A Thermodynamic Treatment of Intelligent Systems." *Rebooting Computing (ICRC), 2017 IEEE International Conference on*. IEEE, 2017.

14 Non-equilibrium Fluctuation Theorems –Intuitive Interpretation

High Dissipation Condition
"The Novice"

Low Dissipation Condition
"The Expert"

- Energy Loss
- Fluctuation
- Adaptation
- Learning

"Thermodynamic Evolution"

- Energy Efficiency
- Homeostasis
- Inference
- Prediction

THERMODYNAMIC COMPUTING SYSTEMS

15   Thermo-dynamic Computing –System Concept

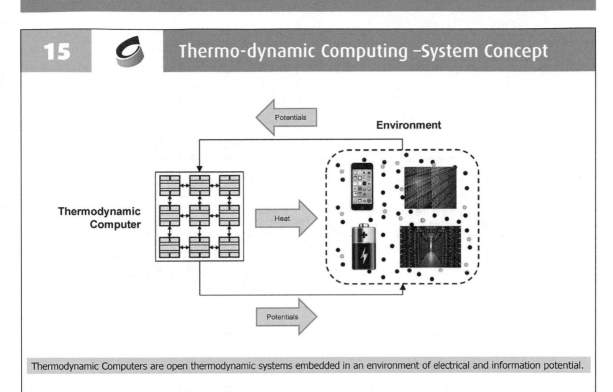

Thermodynamic Computers are open thermodynamic systems embedded in an environment of electrical and information potential.

16   Thermo-dynamic Computing -Basic Architecture

- A generic fabric of thermodynamically *evolvable cores* embedded in a *reconfigurable network* of connections.

- Energy is the "language" of the network

- *The job of the entire system – network and cores - is to move energy from inputs to outputs with minimal loss.*

- Losses within the TDC create variations (fluctuation-dissipation)

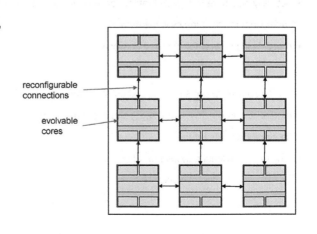

17 **Thermo-dynamic Computing -Conceptual Illustration**

- Networks of simple ECs form a larger Thermodynamic Computer (TDC)

- The "problem" is defined by the energy / information potential in the environment.

- Programmers can fix some of the ECs to define constraints / algorithms that are known to be of value.

- Dissipation within the network creates fluctuations over many length and time scales and thereby "search" for solutions over a very large state space.

- Structure precipitates out of the fluctuating state and entropy production increases in the environment as free energy flows through the network and dissipation decreases

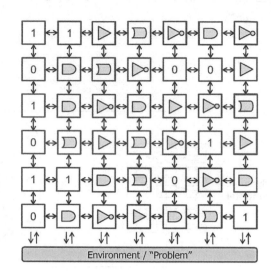

18 **Representations of Energy**

- Explicit representations – e.g. analog signals / pulses

  - naturally distribute energy and satisfy conservation laws (for energy, charge, etc.)

  - suffer from resistive losses in the network

  - probably impractical for the large networks

- Implicit representations – e.g. "spikes" or "messages" or "numbers"

  - require coding (energy → message) and decoding (message → energy) at the cores

  - do not suffer resistive losses

  - require an accounting system to satisfy conservation laws

    - analogous to money in economic systems

energy flow

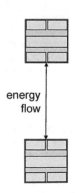

19 — Thermodynamic "Bit"

Circuit Concept

- Unstable state / inherent variations
- Environment influences variation / selection
- Selected state feeds back to the environment
- Weighting changes relative influence

Thermodynamics

The Thermodynamic Bit is a simple evolvable element.

24

20 — Thermodynamic Neural Network

- Nearest neighbor 2D grid with periodic boundary conditions and connection / domain restrictions
- Each node states on [-1,1]
- Adaptive weights among connected nodes
- Driven by 8 pairs of periodic ±1 nodes
- Adaptation from thermodynamic relaxation
- Losses from thermodynamic fluctuation
- Conservation laws strictly enforced
- Naturally recurrent
- No
    - Back-propagation
    - Learning rates
    - Decay rates

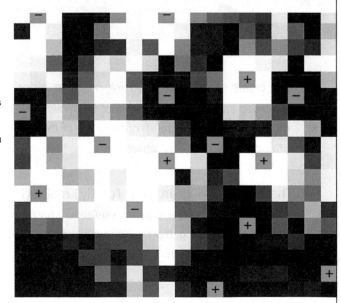

CONCLUSIONS

21 Summary of Ideas

- *Thermodynamic evolution* is the idea that thermodynamic potential in the environment drives the creation and maintenance of organizations to relieve that potential by increasing the thermodynamic entropy in the greater environment.

- *Evolution is realized via thermodynamic self-consistency* - imperfect energy transport yields energy dissipation resulting in fluctuations of configuration that may be stabilized if energy transport is improved.
  - Dissipation drives adaptation
  - Thermodynamic efficiency implies the ability to predict future inputs.

- A *Thermodynamic Computer* is a system that evolves its organization to transport energy in response to electrical and information potential in its environment.
  - *The evolution* of a TDC can be biased through *programming, training* and *rewarding*.
  - Transport losses can be mitigated via *encodings of energy coupled with an accounting system* that maintains energy conservation.

22 Where are we now

- The goals of my 2010 DARPA program titled "Physical Intelligence" were
  - Develop an abstraction / theory of thermodynamic evolution
  - Create a model system(s) for the study of thermodynamic evolution
  - Create simple demonstrations of thermodynamic evolution

- These three goals are now (partly) realized
  - Abstraction / theory – e.g. non-equilibrium fluctuation-dissipation theorems
  - Model system – e.g. thermodynamic neural network
  - Simple demonstrations – e.g. arbortrons

23   Thermodynamic Computing Workshop

24   Why is this so hard?

- The ideas presented here
  - Contradict the prevailing *mechanistic, cause-and-effect, brain-as-a-machine, computing paradigm.*
  - Contradict the current computing practice *where the systems are designed to prevent evolution.*
  - Cannot be described within the current computing paradigm.
  - Imply that *causation is the product of evolution*, thereby *contradicting prevailing scientific and religious doctrine.*
  - Imply that the *universe is uncaused* and *evolved from a thermodynamic singularity that requires no precedent.*

- We have the cart before the horse because we don't know what the horse is.
  - *The horse is thermodynamic evolution.*
  - *The cart is all the dazzling stuff that it creates.*

- We are trapped by unresolved paradoxes
  - The *First Cause* paradox of philosophy
  - The *Reductionist* paradox of physics
  - The *Where Do Algorithms Come From* paradox of computing

- *We are stuck in socio-cultural-technological-scientific local minimum*

25 What Is Intelligence?

Intelligence is the Thermodynamic Evolution of Organization

- *Thermodynamic Evolution means*
  - variation through dissipation,
  - selection by relief of thermodynamic potential / creation of entropy in the larger environment
- *Organizations are*
  - meta-stable, spatio-temporal structures
  - the artifacts of thermodynamic evolution
  - constraints on future thermodynamic evolution
  - sources of thermodynamic potential for future thermodynamic evolution
  - algorithms are a type of thermodynamically evolved organization

The challenge of building intelligent electronic systems is not just the creation of better computers and algorithms, it is the creation of systems that evolve thermodynamically.

CHAPTER 06

Brain-like Cognitive Engineering Systems

Jan Rabaey

University of California

As scaling is coming to a screeching halt, the power and memory walls threaten AI progress based on conventional Von Neumann architectures. In this chapter, Jan Rabaey describes how the fundamental architecture of biological neural systems is different from that of conventional computing systems. Biological computational functions are spatially distributed with little or no time multiplexing. There is no differentiation between memory and logic as both of them are intrinsically intertwined. This effectively eliminates the memory-logic bottleneck present in Von Neumann computing and suggests an in-memory computation model. But, Rabaey also argues reverse-engineering the brain without an adequate theory is infeasible. High/hyper-dimensional (HD) computing is proposed as a possible future computing paradigm for AI applications, which based on theoretical underpinnings. This chapter presents research in pursuit of an HD processor spanning multiple layers of abstraction, i.e., theory, algorithms, computer architecture, and emerging devices.

1 Brain-like Cognitive Engineering Systems

Technology scaling and advancing device technologies have played a major role in making computational engines continuously more efficient. Yet that efficiency is still a couple of orders of magnitude away from what the human brain is capable of. Bridging that gap using traditional models and techniques is becoming increasingly harder due to implicit limitations and/or bounds in the devices, architectures and computational paradigms. The main question to ask is if computational techniques inspired by our current understanding of how the brain functions could help to overcome some if not most of these limitations. In this paper, we explore a number of the properties of the brain function, and how these can/may map into emerging nanotechnologies. Just to name a few these properties: approximate pattern-based computation; close intertwining of logic and memory; 3D topology; learning-based programming model; sparsity and function-specific mapping. How these properties could translate into actual and efficient realizations is illustrated with a set of concrete examples using a hyper-dimensional computing approach.

2 It's Time To Rethink Computing

Computing has a truly profound impact on society, and has changed how we work, operate, live and interact. Yet, over the more than 7 decades since Alan Turing and John Von Neumann laid out the basic concepts of computing little has changed. Virtually all computing performed today is based on the instruction-set architecture model that they brought to life. Yet there are many reasons to believe that this is about to change, and that novel models of computation are about to emerge.

2   **It's Time To Rethink Computing**

IT'S TIME
TO RETHINK
COMPUTING

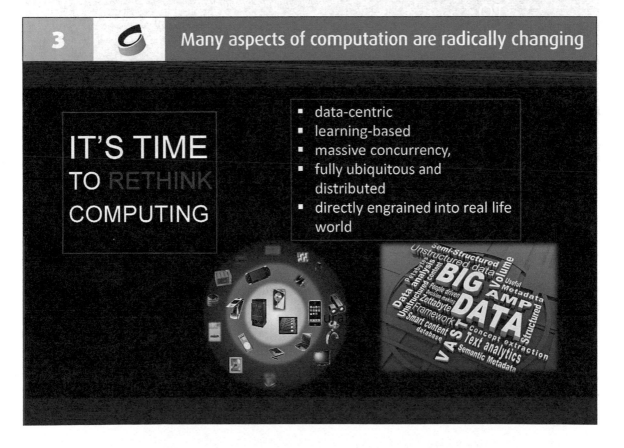

3   **Many aspects of computation are radically changing**

IT'S TIME
TO RETHINK
COMPUTING

- data-centric
- learning-based
- massive concurrency,
- fully ubiquitous and
  distributed
- directly engrained into real life
  world

3   Many aspects of computation are radically changing

The first reason is that the nature of what computing itself means is changing. Over the long history of computing, the focus primarily has been on the essence of computation itself, and on how to maximize the performance of the machines we construct. With electronic interfaces becoming truly ubiquitous and the whole world gradually becoming digitized, the emphasis is moving on how to process, make sense and act on all the data that is being produced. Data is moving center stage. This has some fundamental impact on how computation itself is perceived and organized. In contrast to the earlier eras, it has become truly ubiquitous (the world as a computer), distributed and connected. Massive concurrency (something that was hard to get to in the past) is now the de facto rule. And most importantly, we are moving from an instruction-based model to a learning-based paradigm, which is entirely different in terms of requirements, organization and structure.

4   The platform of old is losing its oomph

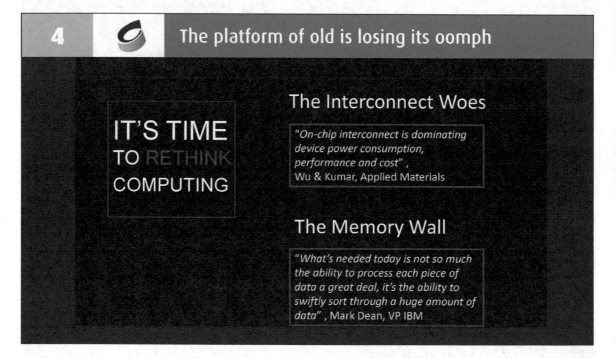

IT'S TIME TO RETHINK COMPUTING

The Interconnect Woes

"On-chip interconnect is dominating device power consumption, performance and cost",
Wu & Kumar, Applied Materials

The Memory Wall

"What's needed today is not so much the ability to process each piece of data a great deal, it's the ability to swiftly sort through a huge amount of data", Mark Dean, VP IBM

While this is happening, a simultaneous transformation in the underlaying technologies is threatening the continuous progress in performance and efficiency of computing that we have become so used to. Moore's and Dennard's Laws, which have been the driving forces behind the amazing run we have witnessed, are losing their oomph and pretty soon will become a minor factor in the advance of the next generation of computational devices. An even more profound challenge to the traditional computational model comes from interconnect.

The prevailing instruction-set architecture model is limited by the bandwidth constraints between data-path and memory, while generations of optimizations and clever architectural design have managed to keep the concept aloft, future improvements are challenged by the fact that interconnect performance and efficiency has not scaled in the same way as transistor performance and by the fundamental separation between memory and computing dictated by the Von-Neumann model.

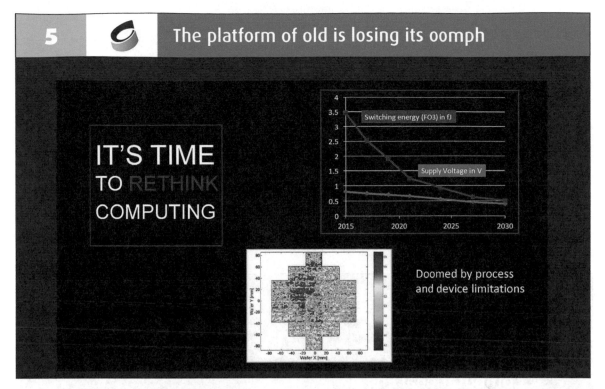

As was mentioned in the previous slide, technology scaling no longer delivers the benefits in performance, efficiency and cost that we have been so used to. More particularly in terms of energy efficiency, voltage scaling has all but stopped. Main hurdles are transistor leakage and process variability. While device and process innovations can make help, it is essential to explore computational technologies and models that can surmount and even thrive under those conditions.

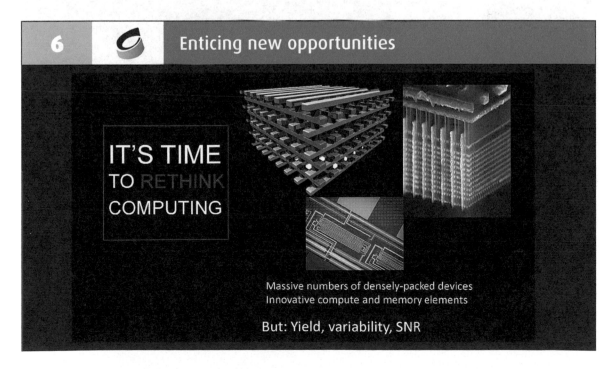

6   Enticing new opportunities

On a positive note, advancements in nanoscale semiconductor devices and systems are continuing unabashedly. Innovative devices for both logic and memory are approaching the nanometer dimension. Even better, the potential of stacking them in multiple layers approaching true 3D integration is becoming a reality. This will allow us to continue to scale density for quite a number of generations. However, those novel platforms come with their own rules and constraints. Assuming 100%

yield in a 3D device with close to 1 Trillion devices does not make sense. Assuming conformity between all these devices is a non-starter as well, and signal-to-noise ratio should be small for energy-efficiency and heat-management purposes. Hence the quest is for computational models that can thrive under those conditions while addressing the system-level constraints expressed in the previous slides. It is time to explore different horizons, our natural environment being one of them.

7   The Neuroscience Promise

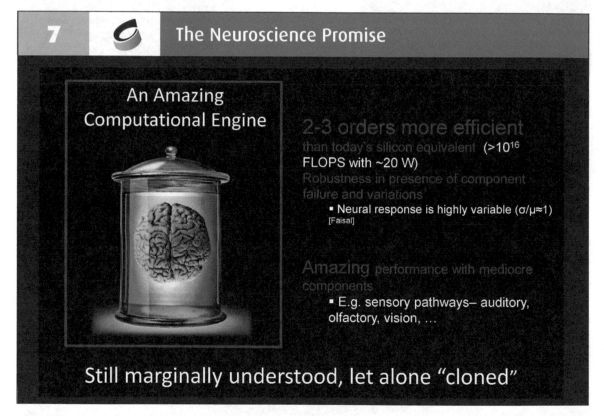

Buried within our natural environment are alternative computational paradigms that are fundamentally different from the Turing/Von Neumann models, and yet have proven their efficacy, efficiency, and robustness over thousands of years. DNA and its related mechanisms have been at the core of the computational machinery that has instigated and propagated life. Similarly the (mammalian) brain

has proven to be an amazing natural computer that surpasses any known semiconductor computer for a broad range of cognitive tasks in terms of efficiency. Some of the other properties of the brain demand special attention as they differ fundamentally from our digital computers.: (1)The neuronal brain exhibits exceptional robustness in the presence of failure and degradation. Neurons fail at a rate of 1000/day. A

7   The Neuroscience Promise

stroke creates substantially more damage. Yet the brain manages to continue providing service. In the case of a minor stroke, it allows for functions to be remapped to other areas. All of this with devices (neurons) that exhibit a huge spatial and temporal variability; (2) The neurons themselves as a device are extremely mediocre compared to the nanometer transistors of today. In spite of that, our brain outperforms in many ways the capabilities of our current computing devices in terms of resolution, response time and adaptivity. This is for sure the case

for the traditional sensory (seeing, hearing, smell, taste, tactile, proprioception and vestibular) and motor pathways. Amazing as it is, it is sad that we are extremely far away from a full understanding of how it operates and functions. Even further away is insight into the nature of cognition. Fortunately there are some principles about neural computing that have become apparent. Principles that we potentially could use to make more efficient computers in the nanometer era, leading to the concept of brain-inspired computing.

8   Contrasting Visions on Computation (I)

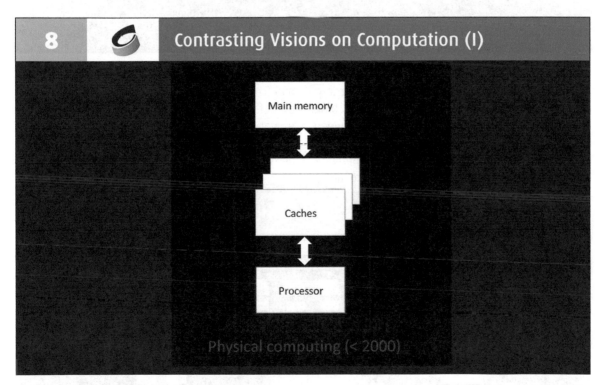

One of the major differentiations between Von-Neumann ("physical") computing is the way how memory and computing are connected. With computing being considered as the most expensive and hard-to-get-at resource that must be optimized at all cost, the majority of research in the computer architecture community has been on how to keep

the processors busy. This has led to complicated hierarchical caching schemes attempting to match memory bandwidth and latency to the performance needs of the CPU. Scaling this has become increasingly problematic especially in light of the ever-larger amounts of parallelism in the processor.

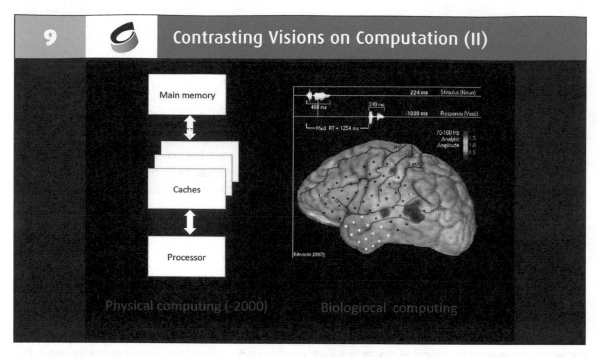

The basic architecture of the neural system is entirely different. Computational functions are spatially distributed with little or no time multiplexing. There is no differentiation between memory and logic as both of them are intrinsically intertwined. This effectively eliminates the memory-logic bottleneck present in Von Neumann computing.

A final aspect of neural computation is the "mostly dark" operational model. Brain regions are inactive by default, and only get turned on when the right excitation patterns are present. This translates into an event-driven execution model reducing the power demands.

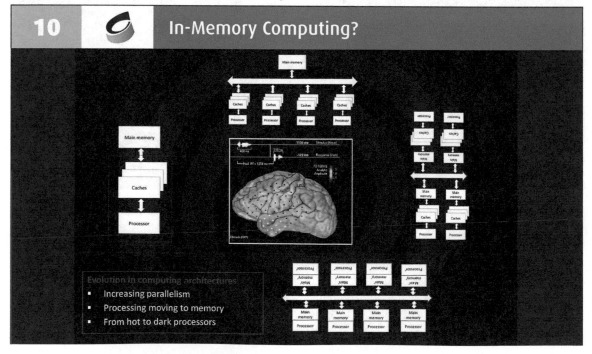

10 In-Memory Computing?

One interesting observation is that the gap between the two approaches has been gradually decreasing over the past few decades. To deal with power dissipation and performance constraints, traditional processors have been embracing increasing amounts of parallelism and specialization, while memory and logic have become more and more intermixed. This has led to the emergence of interesting concepts such as in-memory and near-memory computing. To follow the example of the neural computing, one can foresee that processors will become mostly dark and be only turned on when the associated data is triggered.

11 There is no such thing as "the brain" (I)

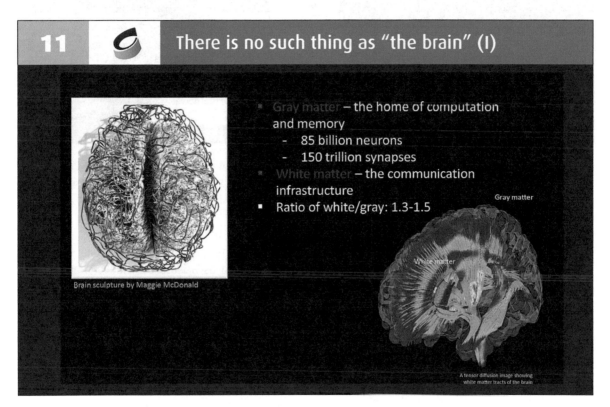

Gray matter – the home of computation and memory
- 85 billion neurons
- 150 trillion synapses

White matter – the communication infrastructure

- Ratio of white/gray: 1.3-1.5

Gray matter

White matter

Brain sculpture by Maggie McDonald

A tensor diffusion image showing white matter tracts of the brain

When considering brain-inspired computing, one should be aware that the brain is not a uniform fabric, and that the way data is represented and computation is performed varies between regions. First of all, neural activity is divided between the central and the peripheral neural systems. The central neural system itself can be divided between gray matter, which is the home of computation and memory, and white matter, which provides the communication infrastructure between the different regions of the brain. One striking observation is that the ratio between white and gray matter is approximately constant, independent of the size of the brain. This holds over a huge range of brain sizes from the smallest mouse to the largest whale. While this ratio seems large, it is small in contrast to the contemporary integrated circuit. In the latter, computing devices are positioned in a very thin layer of only a few um, while the more than 10 layers of interconnect occupy a volume that is orders of magnitude larger. A major differentiator between the two is that the topology of the brain is fully three-dimensional, which leads to substantial density gains.

12 There is no such thing as "the brain" (II)

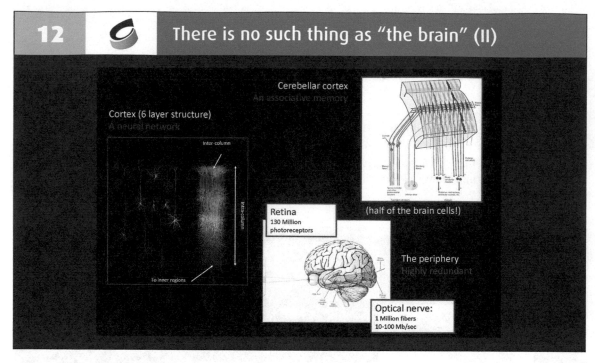

Even within the gray matter major differences exist in terms of composition and interconnect topology between the various regions. While the neuron is the atomic building block, it exists in many forms and shapes. Let us look first at one example of the organization in the periphery, that is the vision system. The retina consists of an array of 130 million photoreceptors, which is substantially larger than the imagers in the best digital cameras available today (which count up to 14 MegaPixels). This provides a huge amount of redundancy, allowing the individual photoreceptors to be highly variant. The retina itself consists of multiple layers of neurons that extract features from the observed image streams and perform a major data compression. This compression is crucial as the data link between the eye and the visual cortex has a limited bandwidth estimated to be between 10 and 100 Megabit/sec. Worth observing is that the data, once reaching the visual cortex, is expanded again by large ration leading to over-redundant representations. A similar model holds for other peripheral sensory systems such as olfaction.

Alternatively consider the cerebellar cortex, which receives information from many parts of the body and the brain and enables accurate and well-coordinated movements. It has a uniform structure that has challenged researchers for years. On closer

inspection, reveals a structure that amazingly resembles a semiconductor memory. It contains a decoder-like structure made out of "granular cells", parallel fibers acting as word-lines, Perkinje cells as memory cells, and their axons functioning as bit-lines. It has been speculated that the cerebellar cortex can be considered as a large associative memory that learns to translate course motor commands into fine-tuned motor actions.

Finally, the cerebral cortex (often simply called "the cortex") is the most recent addition to the mammalian brain. It plays a key role in attention, perception, awareness, thought, memory, language, and consciousness, and contains 14 to 16 billion neurons packed in a thin layer of 2-3 mm folded around the brain. It is organized in cortical columns consisting of 6 layers of neurons. The outer neurons support inter-columnar communication while the inner ones connect to the other brain regions. It is this architecture that has inspired most of the neural network model.

This discussion only touches a few regions of the brain, and many other sub-systems *and their relationships) are under exploration today. The bottom line is that the brain can inspire many different types of computational structures and models.

13 Common Properties of Neural Computing

Computational model
- Learning-based programming paradigm
- Approximate (statistical) & mostly analog
- Data represented in many ways
  - Patterns, phase relations, distributions
- Randomness as a feature

Structure and topology
- Function mapped to space
  - no time multiplexing
- Intertwined memory and logic
- Embarrassingly parallel
- Sparse

One thing is for sure though: our understanding of the brain and its operational model is still very primitive. Advanced imaging techniques, passive and pro-active brain interfaces, and complex topology models of the brain have accelerated our insights over the past decade. Yet even so, we are still very far away from a full understanding of what transpires under the hood. Notwithstanding the predictions of a "singularity" to occur about 20 years from now, reality seems to dictate that a full clone of human brain will likely take far more than that.

There are however a number of properties of neural computing that we have learned, and that highlight the huge gap between Von-Neuman physical computing and brain-like biological computing. A reflection on those properties can help to open the door for a systematic and methodological approach to brain-inspired computing. They can be divided into two classes: (1) computational model and data representation; and (2) structural and topological.

14 Reverse-engineering the brain

One truly important word of caution regarding brain-inspired computing must be raised at this point. There exists a long standing belief that a simulacrum of the brain, built from the ground up from neuron models and their interconnections ("connectomics"), would yield a computational system that behaves and functions just like the organic brain. The truth in the matter is that such a construct will do little at all without an understanding on how the brain topologies came to be and how they operate and evolve. Only then will we be capable of crating scalable, multi-functional engines that possess some of the key characteristics of brain computing, such as reasoning by analogy, generalization from examples, continuous learning and adaptation in response to (dramatic) failure. The field of neuroscience is making rapid progress due to the introduction of advanced tools and methodologies that help to observe the operation of the brain in real-time and with high resolution. Yet there is a long way to go. In an insightful paper, Eric Jones and Konrad Kording [PLOS January 2017] mused how much we would understand about the operation of a microprocessor if the observations we would have would be the average voltage/energy over a module or a set of at most 100 voltage measurement points. Without the concepts of Boolean algebra and instruction-set computing, that would be very little indeed.

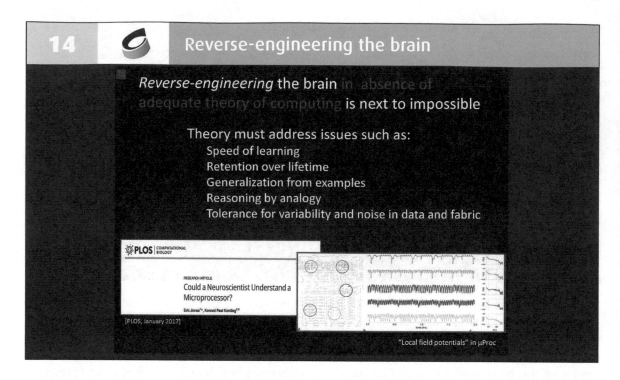

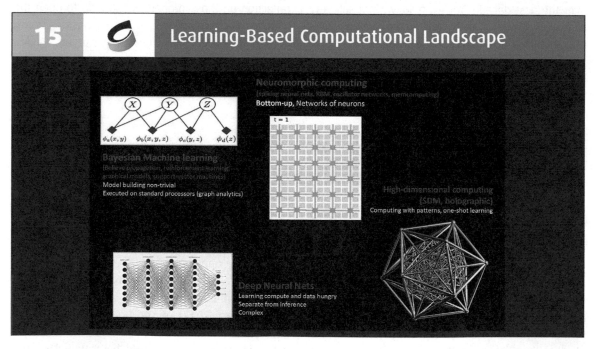

As stated earlier, the main differentiator between neuro-inspired and Turing/Von Neumann-based computational approaches is the learning versus algorithm-based programming models. However, learning-based does not necessarily mean neuro-inspired. The field of Bayesian machine-learning (B-ML) has delivered a broad range of excellent techniques and approaches to analyze and reason about large data sets by building on the extensive bag of tools offered by statistics. Observe that virtually all of the B-ML tools run on traditional processors (CPU/GPU) and do not exploit the fact that the

15 Learning-Based Computational Landscape

data is statistical in nature. On the other end of the spectrum, neuromorphic computers aspire to match the computational structure of the brain both in the data-representation (spike-based) and the topology (networks of neurons). In between are the neuro-inspired approaches that borrow some properties of brain-computing, but translate them into a novel and unique computational model. Deep neural nets are an excellent example of such. The computational topology (layers of connected neurons) is borrowed directly from the brain, but the learning/inference mechanisms are not. The resulting structure has proven to be extremely effective for classification tasks when plenty of training data is at hand. Similarly, high-dimensional computation builds on the pattern-based data representation in the brain as well as the associative memory function of the cerebellum, but layers a unique computational algebra on top of this. Both approaches hence display properties that are similar to neuron-based computing (such as statistical operation and fault-tolerance).

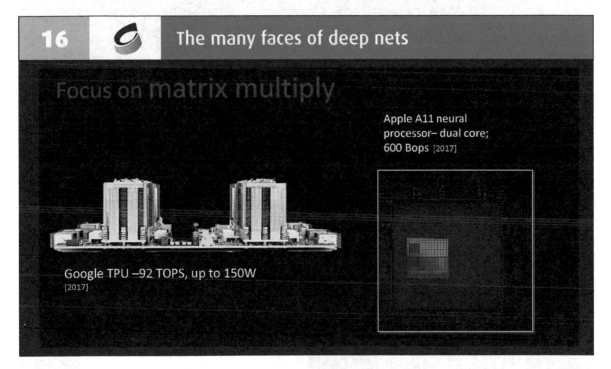

16 The many faces of deep nets

Focus on matrix multiply

Apple A11 neural processor– dual core; 600 Bops [2017]

Google TPU –92 TOPS, up to 150W
[2017]

To evaluate how important these properties are and how they can lead to innovation at all levels of the design hierarchy, let us explore the realization of an energy-efficient deep-net processor. A quick back-of-the-envelope shows that vector-matrix multiplies make up most of the computational complexity to be provided by the processor. Hence, adding a matrix-multiply systolic accelerator to CPU/GPU core makes more than sense, an approach that was adopted by the Google TPU and many other deep net processors.

Yet, this not necessarily address the energy-efficiency issue with power still topping 150W. Mobile processors solved this problem by creating a dedicated and customized accelerator processor primarily focusing on the inference part, and leaving the learning part to off-line processing. An example of such is the neural accelerator imbedded into the Apple A11 processor. To push power consumption lower however requires innovation exploiting the specific properties of neural computation.

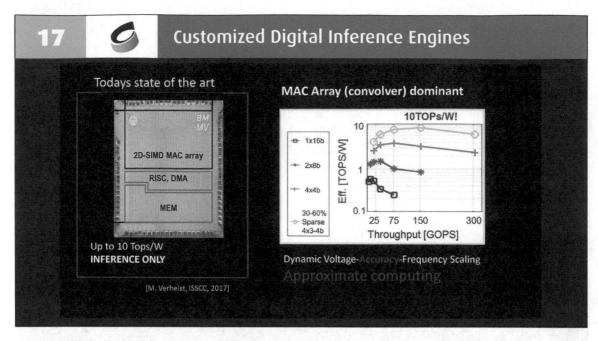

17   Customized Digital Inference Engines

Todays state of the art

2D-SIMD MAC array

RISC, DMA

MEM

Up to 10 Tops/W
INFERENCE ONLY

[M. Verhelst, ISSCC, 2017]

MAC Array (convolver) dominant

10TOPs/W!

- 1x16b
- 2x8b
- 4x4b

30-60%
Sparse
4x3-4b

Dynamic Voltage-Accuracy-Frequency Scaling
Approximate computing

For instance, it is possible to exploit the statistical nature of the neuro-inspired processing to configure the accuracy in the matrix-multiply engine to trade-off computational performance versus classification accuracy. This adds a new "accuracy" button to the arsenal of tools of the energy optimizer (in addition to the well-known voltage and frequency scaling ones). The key lesson learned here is that the approximate computing approach nicely fits this class of problems.

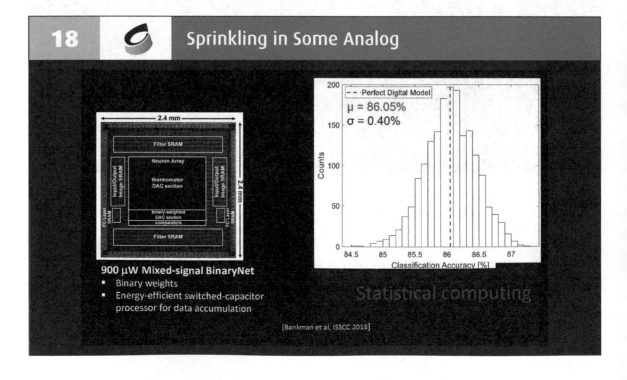

18   Sprinkling in Some Analog

2.4 mm

Filter SRAM

Neuron Array

thermometer
DAC section

binary-weighted
DAC section
comparators

Filter SRAM

900 µW Mixed-signal BinaryNet
- Binary weights
- Energy-efficient switched-capacitor processor for data accumulation

$\mu = 86.05\%$
$\sigma = 0.40\%$

Perfect Digital Model

Statistical computing

[Bankman et al, ISSCC 2018]

18 Sprinkling in Some Analog

Once we accept the approximate computing concept, the next step comes quite naturally. Why not perform the multiplications in the analog domain, similar to what happens in the biological synapse? While not entirely trivial (how to store the weights, etc), this approach allows for another major improvement in energy efficiency. However since the result now longer is deterministic, the computation has become statistical, and spatial and temporal variations of devices come into play. Careful tuning of the technology as well as design centering approaches are necessary to yield acceptable results.

19 In-Memory COMPUTING

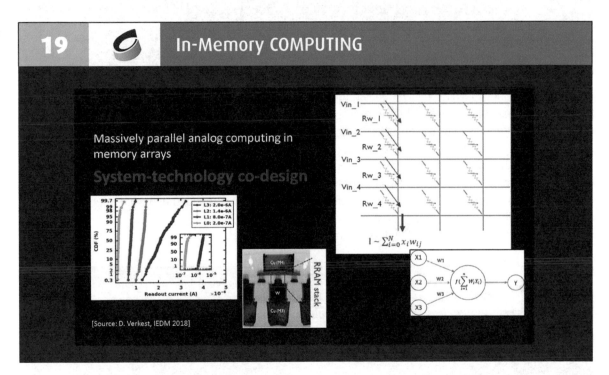

The various optimizations discussed in the previous slides (observe that this is only a sub-set of the techniques that deep-net designers have come up with) reduced the power dissipated in the vector-multiply array to the point that it is no longer dominant. Fetching the data and the weights from the surrounding memories is now eating the lion share of the budget. With brain-like computing in mind, the next step is obvious: merge the data storage and the computation. This translates into massively parallel analog in-memory computing. This now requires not only innovation at the architecture and the circuit but also at the technology level, and requires a design methodology that crosses all levels of the design hierarchy.

Hopefully this short discussion demonstrates that the introduction of more and more neuro-inspired concepts leads can lead to substantial improvements in a number of design metrics. At the same time, it requires fundamental changes in the design flow and requires a system-technology co-optimization approach that is just in its infancy.

20 High-Dimensional Computing as an alternative

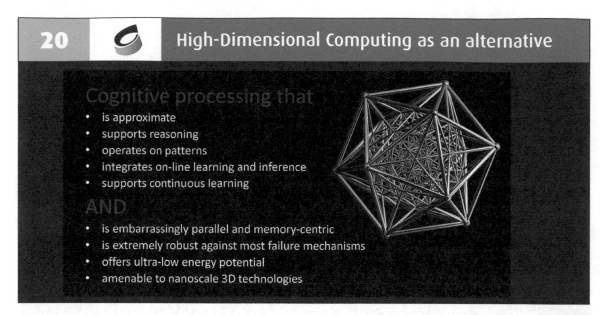

Cognitive processing that

- is approximate
- supports reasoning
- operates on patterns
- integrates on-line learning and inference
- supports continuous learning

AND

- is embarrassingly parallel and memory-centric
- is extremely robust against most failure mechanisms
- offers ultra-low energy potential
- amenable to nanoscale 3D technologies

Deep neural nets (while wildly popular today) are just one of the possible neuro-inspired computational models. To its disadvantage, the computational requirements for learning and inference are substantially different, with the former being more complex as it exploits gradient descent optimization. Also training the huge number of weights (often in the millions) requires a tremendous amount of data. Finally, there is still limited understanding of the computational concepts governing its operation, translating to little transparency and sometimes unexpected or unwanted results

For a number of tasks, fast and continuous learning is of essence, often based on little data. Hence other approaches may be more applicable. The high-dimensional computing approach offers just such an alternative.

21 Hyperdimensional Vectors – the Concept (I)

21   Hyperdimensional Vectors – the Concept (I)

The picture in this slide presents an intuitive understanding of computation in high-dimensional spaces. Just as the space surrounding us, high-dimensional spaces are naturally sparse. Related objects are clustered, that is they are proximate (given a distance measure).

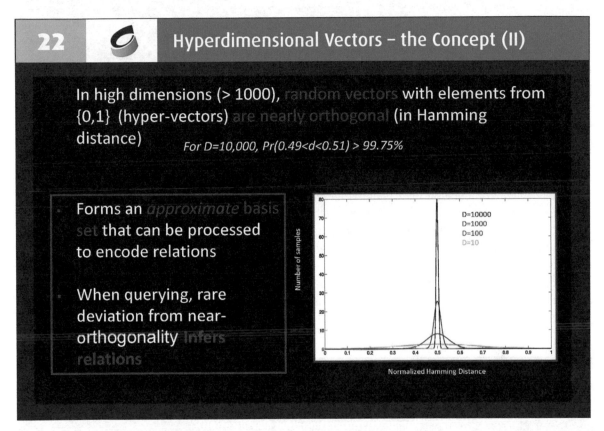

In high dimensions (> 1000), random vectors with elements from {0,1} (hyper-vectors) are nearly orthogonal (in Hamming distance) *For D=10,000, Pr(0.49<d<0.51) > 99.75%*

- Forms an *approximate* basis set that can be processed to encode relations

- When querying, rare deviation from near-orthogonality infers relations

One way to create a high-dimensional space is to use high-dimensional (D> 1000) random vectors of symbols as the basis set. In the simplest form, these symbols could be binary (from the {0,1} or {-1,1} set) or could be integers or even complex numbers.

Assuming the {0,1} set, it is easy to see that two vectors of randomly chosen vectors ("patterns") are pseudo-orthogonal. In fact, the Hamming in distance between two such vectors follows a binomial distribution. As shown in the plot, if the vector dimension D is high enough, the probably of the distance to be 0.5 (as normalized over the dimension) approaches 100%.

In high-dimensional computing, the basis set will consist of a number of randomly chosen vectors. Using a set of arithmetic operations, these will be processed and combined to encode relations. A guiding principle is that any two vectors that deviate from near-orthogonality must have some form of established relation.

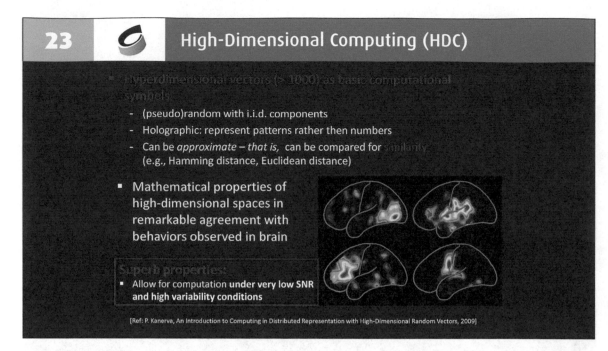

23 — High-Dimensional Computing (HDC)

- Hyperdimensional vectors (> 1000) as basic computational symbols:
  - (pseudo)random with i.i.d. components
  - Holographic: represent patterns rather then numbers
  - Can be *approximate – that is,* can be compared for similarity (e.g., Hamming distance, Euclidean distance)

- Mathematical properties of high-dimensional spaces in remarkable agreement with behaviors observed in brain

Superb properties:
- Allow for computation **under very low SNR and high variability conditions**

[Ref: P. Kanerva, An Introduction to Computing in Distributed Representation with High-Dimensional Random Vectors, 2009]

The concept of computation with patterns rather than numbers is in remarkable agreement with behaviors observed in the brain, in which collections of neurons only fire when presented with selected patterns of input signals.

Patterns do not have to be exact but can be approximate. This leads to a unique property in terms of robustness in presence of errors and variations. Bit-errors matter little due to the natural redundancy in the representation. The proportion of the allowable number of errors increases with dimensionality. This is in contrast with the traditional binary number representation of data, where a single bit error can significantly change its meaning. HDC hence opens the door for computation at low SNR level, and higher energy-efficiency.

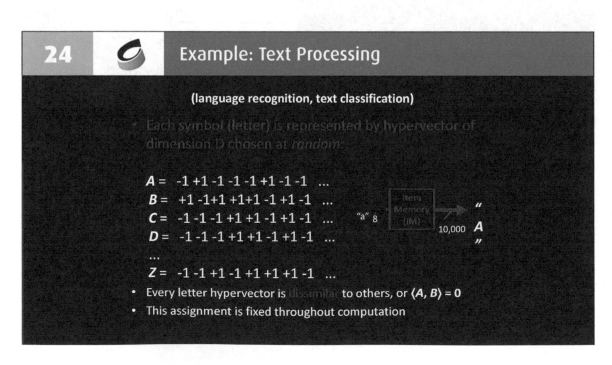

24 — Example: Text Processing

(language recognition, text classification)

- Each symbol (letter) is represented by hypervector of dimension D chosen at *random*:

$$A = \ -1\ +1\ -1\ -1\ -1\ +1\ -1\ -1\ \ ...$$
$$B = \ +1\ -1+1\ +1+1\ -1\ +1\ -1\ \ ...$$
$$C = \ -1\ -1\ -1\ +1\ +1\ -1\ +1\ -1\ \ ...$$
$$D = \ -1\ -1\ -1\ +1\ +1\ -1\ +1\ -1\ \ ...$$
$$...$$
$$Z = \ -1\ -1\ +1\ -1\ +1\ +1\ +1\ -1\ \ ...$$

"a" 8 → Item Memory (iM) → $10{,}000$ " A "

- Every letter hypervector is dissimilar to others, or $\langle A, B \rangle = 0$
- This assignment is fixed throughout computation

24 Example: Text Processing

As a simple example of the basic concepts of HD processing, consider the problem of recognizing the language a fragment of text is written in. Rather than trying to understand the semantics of the language, it suffices to observe the occurrence of typical patterns. One may consider a piece of text as a stream of characters. For all Indo-European, a set of 26 possible characters suffices. Rather than representing each character as a number (such as ASCII), HD maps each into a high-dimensional space with a random hyper-vector chosen for each character, All "character" vectors are pseudo-orthogonal. Many different ways of performing the mapping can be envisioned. The simplest one would store a number of vectors into a memory, and to use the value of the input signal as the index into the memory.

25 Computing a Profile Using HD Arithmetic

- Addition (+) is good for representing sets (bundling), since sum vector is similar to its constituent vectors.

  - ○ ⟨A+B, A⟩=0.5

- Multiplication(*) is good for binding, since product vector is dissimilar to its constituent vectors.

  - ○ ⟨A*B, A⟩=0

- Permutation (ρ) makes a dissimilar vector by rotating, and is good for representing sequences.

  - ○ ⟨A, ρA⟩=0

- * and ρ are invertible and preserve the distance

The structure of the data is captured by performing a well-defined set of operations on the random seeds. These operations are used for encoding and decoding patterns. The power and versatility of ordinary arithmetic derives from the fact that addition and multiplication form an algebraic field.

The addition and multiplication operations on vectors likewise form, or approximate, an algebraic field. High-D vector algebra has two more operations. Multiplication by scalar puts it on par with linear algebra, and permutation of vector components takes it beyond both arithmetic and linear algebra.

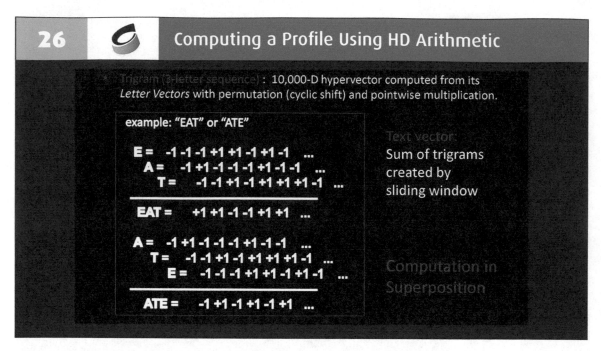

These operations can now be used to capture the structure of the incoming data. For example, a unique representation for a trigram, that is, a sequence of three characters, can be constructed by permutation and multiplication. E.g. the vector representing "EAT" is formed as follows: "EAT" = $\rho 2$("E"). ρ("A")."T". The complete text can then be encoded by sliding a three-character window through the text, and adding the encoded trigrams. The addition superimposes all the trigram vectors into a single vector representing the complete text. The addition (superposition) ensures that the resulting vector is somewhat "similar" to the composing elements.

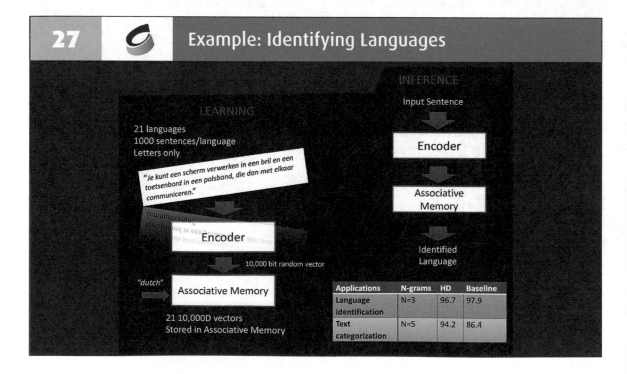

27 Example: Identifying Languages

The language recognition then proceeds as follows. During learning, a number of sentences are fed through the encoder ultimately resulting in a single vector representing all of the absorbed text in that language. The result is stored in an associative memory, and annotated with the proper language label. During inference, exactly the same sequence is performed with the exception of the last step. Rather than storing the resulting text vector in the memory, a search is performed for the closest match. The distance metric could be as simple as a Hamming distance. The label of the vector with the closest match is returned as the result. If no close match is found, the "unknown" label is returned. Notice that the associative memory is truly at the core of the approximate computing concept of HDC. Even with this simple scheme, the results for language identification and text categorization are excellent and au par if not better than the state of the art.

28 HD Processing – An Example

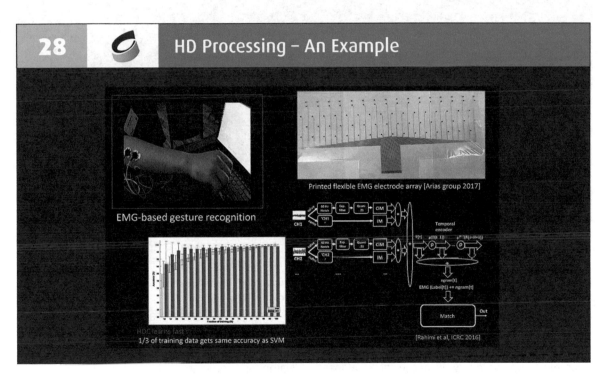

The same concept can be used to perform classification or event detection in a broad range of streaming sensor data applications. In the example shown here, gesture recognition is performed using a set of EMG (Electromyography) sensors placed on the arm. The difference with the previous example is that multiple input channels are combined. The encoder combines both the spatial ("which channel is the sample from") and the temporal ("at what time was the data collected") into a single vector. For a given sample time, the collected data for a given channel is mapped into the HD space, and multiplied by the HD code for the channel forming a 'Key-value" pair. Summing all the channels leads to a vector representing a single sample time ("space"). These vectors are then combined in time using an n-gram encoder, similar to the one used in the text example. The results for even large numbers of gestures (> 21) are substantially better than the state of the art. More importantly, new gestures can be learned very quickly, and give high accuracy results with far less training data than other approaches such as support-vector machines (SVM).

29 · From Classification to Reasoning

What is the Dollar of Mexico?

Learning

R1 = Country * USA + MoneyUnit * Dollar + Population * 320M + ...
R2 = Country * Mexico + MoneyUnit * Peso + Population * 120M + ...

...

Queries

<R2 * Country> = Mexico
<R2 * <R1 * Dollar> > = Peso

...

- Yellow color codes components operating with HD distributed representation.
- <> stand for associative match

The previous results, demonstrating simple topology and quick on-line learning, are interesting for sure, yet do not really set HD apart from other learning-based techniques. The fact that HD is underlaid by an algebra adds another dimension to this discussion. Once again, we will use a simple example to illustrate the potential power of this. Let us consider a simple database query. In a common database, various pieces of information about an object would be stored as a list of key-value pairs. For instance, a database entry for a country such as the United States would consists of a number of fields with keys such as the country name, its monetary unit, its population, etc. To find the value for any field would require a search for appropriate key and its value. In an HD representation, the total database entry is a single vector, consisting of a superposition ("+") of the different fields, each obtained as the product of the key-value pairs.

Extracting a piece of data such as the name of the country just requires a multiplication of the database entry with the HD label for the Country Name (and some associative memory cleanup). This however allows for some non-intuitive searches to be performed. There is no precise answer for the question: "What is the dollar of Mexico?", yet we as humans intuitively understands what is meant, and a standard data base query will yield little. In HD, this however translates into a simple mathematical operation.

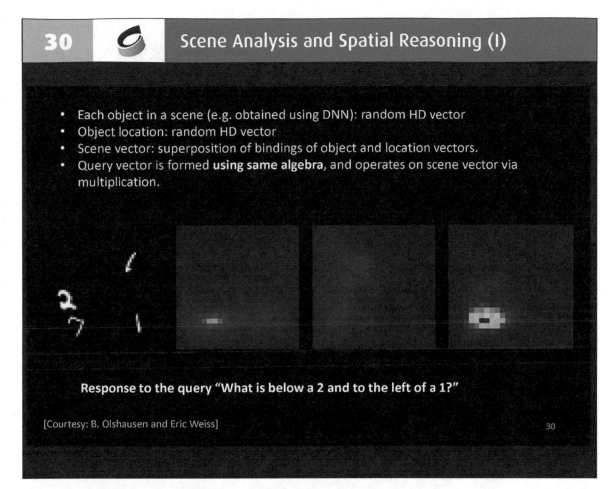

30 · Scene Analysis and Spatial Reasoning (I)

- Each object in a scene (e.g. obtained using DNN): random HD vector
- Object location: random HD vector
- Scene vector: superposition of bindings of object and location vectors.
- Query vector is formed **using same algebra**, and operates on scene vector via multiplication.

Response to the query "What is below a 2 and to the left of a 1?"

[Courtesy: B. Olshausen and Eric Weiss]

That same mechanism can be used to apply HD to the domain of image understanding. Here we see how HD naturally handles the problem of compositionality in a visual scene representation. This is something that is difficult for traditional neural nets to handle. At left is shown a simple scene containing multiple objects (MNIST digits). We wish to obtain a holistic scene representation that allows for random access of information about different parts of the scene, or which can be queried to perform spatial reasoning about the scene. This is accomplished by forming a 'scene vector' that is the superposition of multiple bindings of 'what' (identity) and 'where' (location) information. To answer the question, "what is below a 2?", we first bind the scene vector with the HD representation of '2'. The output immediately provides a vector that encodes its location, and this is then bound with the representation of 'below' to reveal a new locus (panel b). This is done again to answer "what is to the left of a 1?" to reveal another locus (panel c). The intersection of these two loci results in a more localized region (panel d). The corresponding location vector is then bound with the scene vector to reveal the answer, which is vector most closely matched to '7'. Thus, the system can form a flexible, high-dimensional visual scene representation that naturally lends itself to performing high-level reasoning about the scene.

31 Scene Analysis and Spatial Reasoning (II)

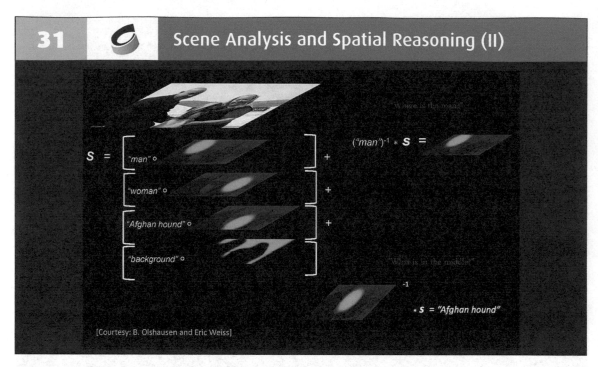

[Courtesy: B. Olshausen and Eric Weiss]

Similar techniques can be used to query a composite image about the placement of objects.

32 HDC maps well into 3D nanostructures

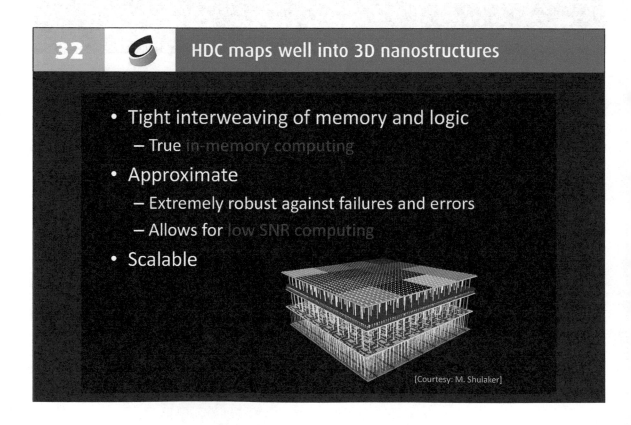

- Tight interweaving of memory and logic
  - True in-memory computing
- Approximate
  - Extremely robust against failures and errors
  - Allows for low SNR computing
- Scalable

[Courtesy: M. Shulaker]

32 HDC maps well into 3D nanostructures

Over the previous slides, we have discussed how computing with patterns similar to what is happening in the brain can lead to some interesting results. The remaining question is if this has also the potential of leading to advances and breakthroughs in energy-efficient implementations. A number of first-order observations are worth making: (i) The length of the vectors (D > 1000) may seem offsetting at first - at least, when compared to the 64-bit word length of todays processors; (ii) at the same time, HD programs are very shallow, requiring very few operations to be performed per sample; and (iii) all HD operations, with the exception of the associative

search, are bit-wise ad require only local connections. However the highest potential for improvement is in the opportunity for system-technology co-design, as discussed earlier. Memory and logic are closely interwoven in the HDC paradigm, opening the door for innovative 3D memory-logic integration. Advanced technologies of this nature often come with yield concerns, but not so for HDC, which can operate robustly at high error rates and low SNR conditions. As such, HDC may be one of the very first computational approaches to have a shot at mimicking the 3D structure of neural computing.

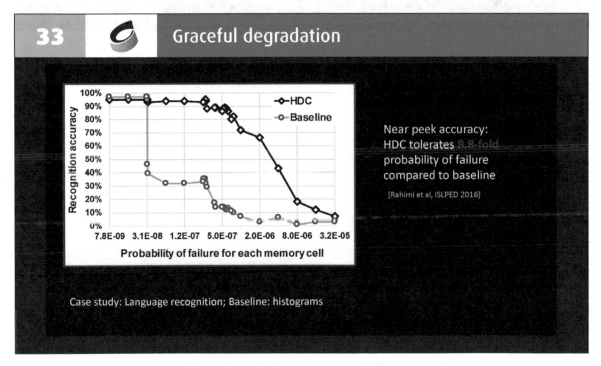

The inherent tolerance of HDC to faults is illustrated by this experiment. The performance of the language recognizer is compared to an approach based on histograms when exposed to an increasing number of errors (bit flips in the memory cells). The processor-based implementation of the histogram algorithm suffers from catastrophic failure once a

small number of errors are introduced. In contrast, the HD implementation degrades gracefully and can operate near- peak for a much larger number of errors. This makes the approach quasi-immune to the unavoidable failures that come with dense 3D integration.

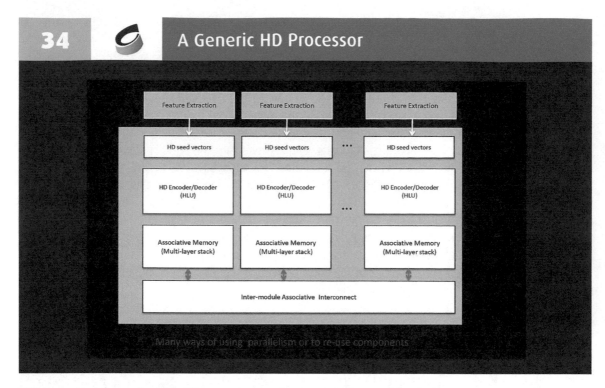

34 A Generic HD Processor

To be concrete, a configurable HD processor capable of running a broad range of applications would look as follows:
- Inputs are either streamed into the processor or are stored in memory.
- In a first step, all inputs and labels are mapped into HD space. This could be with the aid of an Item Memory, pre-storing a collection of pseudo-orthogonal HD vectors, or on-the-fly creation using pseudo-random generators.
- The configurable HD data path encodes the input streams into a single result vector
- Which is either stored (learning) or searched for (inference) in the associative memory

This structure is extremely modular. For instance, it is simple to extend the vector length by chaining multiple chips together. In addition, all connections are only nearest-neighbor.

35 Associative Memory

The most challenging part of the processor realization is the associative memory (AM), whose task it is to find the closest match between an input vector and all the stored vectors. Distance metrics can be simple (such as a Hamming distance) or more complicated (such as a cosine similarity metric). As such, the AM stands for a large fraction of the power budget as well as the latency of the HD processor. Similar to the vector-matrix multiplier unit in the deep net processor, a set of techniques can be used to reduce the imprint of the AM: reduced accuracy, analog processing and in-memory computing.

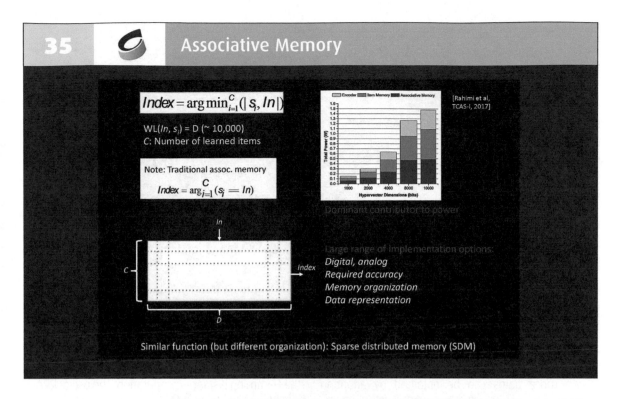

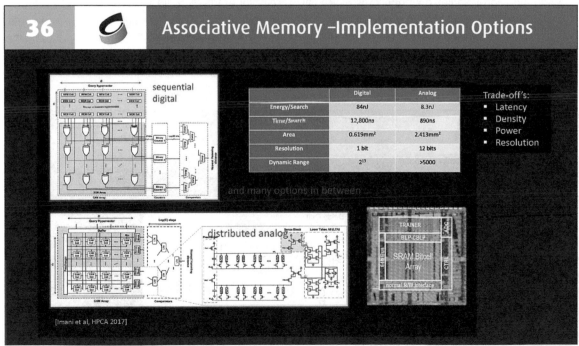

Two possible realizations are compared. In the first approach, the distance-metric and search computations are performed in digital in the periphery of the memory module. In a second approach, distributed analog logic and comparators are used for an in-memory realization. This results in an order-of-magnitude reduction in energy and search time (simulated). This comparison is based traditional SRAM memory cells.

37 Device opportunity: Ferroelectric CAM

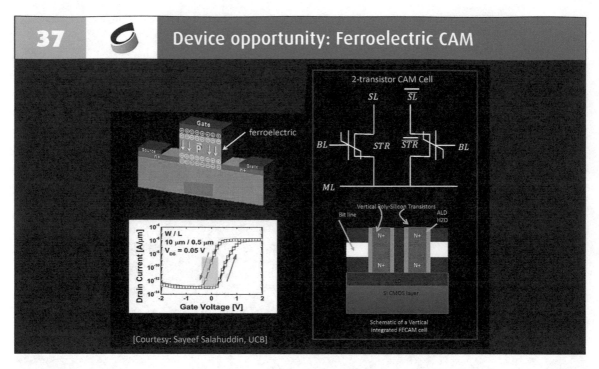

[Courtesy: Sayeef Salahuddin, UCB]

Even more gains can be obtained by adopting emerging non-volatile memory cells, such as PCRAM, MRAM or RRAM. This picture shows how the use of a ferroelectric CAM reduces the total transistor count in the cell to two, while simultaneously implementing the bitwise XOR function needed in the Hamming distance computation. The use of RRAM cells, for example, could lead to even more advances, storing multiple bits in a single cell. While this may lead to more variance, this is compensated for by the highly redundant nature of the HD representation.

38 Device Opportunity: Exploiting Process Randomness

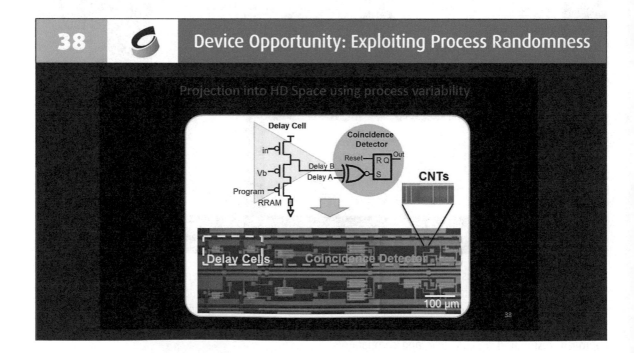

38 Device Opportunity: Exploiting Process Randomness

Other features of emerging technologies can be exploited as well. Earlier we stated that the mapping of input vectors into the HD space can be performed in a lookup (item) memory. Another option is to create the vectors on the fly using a random-number generator. We have demonstrated that the combination of high variability devices such as carbon nanotubes and RRAM cells can lead to generators that have a huge amount of variability (compared to the mean) making them ideal candidates for the mapping. Again, results such as this are a direct result form system-technology co-design.

39 Process Opportunity: 3D integration of RRAM and logic

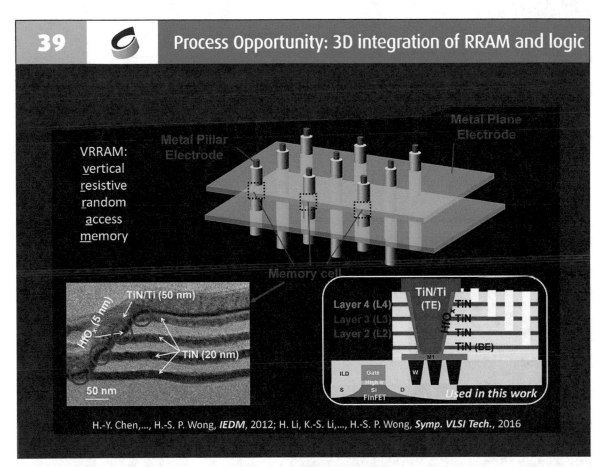

H.-Y. Chen,..., H.-S. P. Wong, *IEDM*, 2012; H. Li, K.-S. Li,..., H.-S. P. Wong, *Symp. VLSI Tech.*, 2016

In collaboration with our colleges in Stanford (Wong, Mitra), we have set out to demonstrate the world's first 3D integrated HD processor prototype, demonstrating the principles outlined before. The process used RRAM, CNT and CMOS logic into a singly 3D integrated entity.

40 HD operations in 3D RRAM

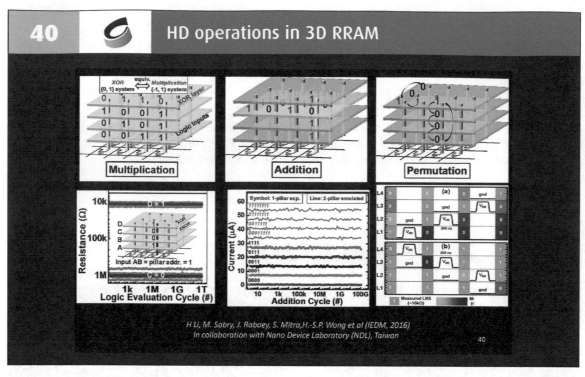

H Li, M. Sabry, J. Rabaey, S. Mitra, H.-S.P. Wong et al (IEDM, 2016)
In collaboration with Nano Device Laboratory (NDL), Taiwan

40

The prototype demonstrated the feasibility of integrating the three HD operations (+, *, <) into the memory. In ISSCC 2018, the same team published the first complete 3D integrated HD processor.

41 Final Reflections: Get inspiration from nature

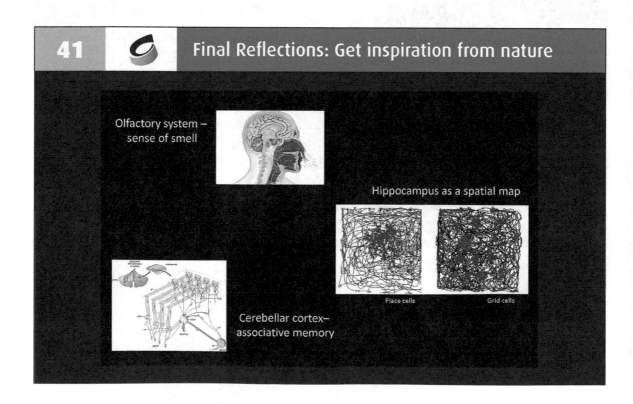

41 Final Reflections: Get inspiration from nature

In summary, it is obvious that there are many opportunities to advance computing beyond the limitations imposed by the computational and technological models of today. There is little doubt that learning-based approaches will continue to expand the horizon of what computation can accomplish. There is much to explore still and nature can serve as a guidance, be it quantum-computing or biological computing. A deeper understanding of how various parts of the brain compute, store, learn and evolve will help in a major way in how to improve the efficacy and efficiency of our next-generation computing devices. A high-level scan of some components of the brain (of which we have at least some understanding) exposes so many different ways on how computation may evolve in the decades to come.

42 Acknowledgments

The many contributions of my students and many of my colleagues to this presentation are gratefully acknowledged.

The support of the the SRC/NSF Enigma Project, the StarNet SONIC Center, Intel Corp., and the member companies of BWRC and SwarmLab is greatly appreciated.

43 Relevant publications

1. P. Kanerva. "Hyperdimensional computing: An introduction to computing in distributed representation with high-dimensional random vectors." *Cognitive Computation*, 1(2):139–159, 2009.

2. Abbas Rahimi, Sohum Datta, Denis Kleyko, Edward Paxon Frady, Bruno Olshausen, Pentti Kanerva, Jan M Rabaey, "High-Dimensional Computing as a Nanoscalable Paradigm," in *IEEE Transactions on Circuits and Systems I* , Issue 99, June 2017.

3. Abbas Rahimi, Pentti Kanerva, José del R Millán, Jan M Rabaey, "Hyperdimensional computing for noninvasive brain–computer interfaces: Blind and one-shot classification of EEG error-related potentials," *10th ACM/EAI International Conference on Bio-inspired Information and Communications Technologies* (BICT), 2017. (Best paper award)

4. Mohsen Imani, Abbas Rahimi, John Hwang, Tajana Rosing, Jan M. Rabaey, "Low-Power Sparse Hyperdimensional Encoder for Language Recognition," in *IEEE Design & Test*, 2017.

5. Abbas Rahimi, Simone Benatti, Pentti Kanerva, Luca Benini, and Jan M. Rabaey, "Hyperdimensional Biosignal Processing: A Case Study for EMG-based Hand Gesture Recognition," in *IEEE International Conference on Rebooting Computing* (ICRC), October 2016.

6. A. Rahimi, P. Kanerva, L. Benini, J. Rabaey, "Efficient Biosignal Processing Using Hyperdimensional Computing: Network Templates for Combined Learning and Classification of ExG Signals," In *Proceedings of the IEEE*, 2018.

7. T. Wu, P.-C. Huang, A. Rahimi, H. Li, M. Shulaker, J. Rabaey, H.-S.P. Wong and S. Mitra, "Brain-inspired computing exploiting carbon nanotube FETs and resistive RAM: Hyperdimensional computing case study," *2018 IEEE International Solid-State Circuits Conference -*(ISSCC), San Francisco, CA, pp. 492-494, 2018.

8. A. Moin, A. Zhou, A. Rahimi, S. Benatti, A. Menon, S. Tamakloe, J. Ting, N. Yamamoto, Y. Khan, F. Burghardt, L. Benini, A. C. Arias, J. Rabaey, "An EMG Gesture Recognition System with Flexible High-Density Sensors and Brain-Inspired High-Dimensional Classifier," in *IEEE International Symposium on Circuits and Systems* (ISCAS), 2018.

9. H. Kim, "HDM: Hyper-Dimensional Modulation for Robust Low-Power Communications," in *IEEE International Conference on Communications* (ICC), 2018.

10. Spencer J. Kent, E. Paxon Frady, Friedrich T. Sommer, Bruno A. Olshausen, "Resonator Circuits for factoring high-dimensional vectors," *arXiv*:1906.11684, July 2019.

CHAPTER 07

BRAINWAY and nano-Abacus architecture

Brain-inspired Cognitive Computing using Energy Efficient Physical Computational Structures, Algorithms and Architecture Co-Design

Andreas G. Andreou

Johns Hopkins University

Continuing on the theme of the prior chapter, Andreas Andreou provides a number of examples of bio-inspired chip designs, many of which are components in systems that solve complex problems of interest to organizations like DARPA. It also describes three waves of computing that match well with ideas presented in chapter 1. This chapter addresses major issues in AI system design, such as data movement, energy limited computation, design methodology, and chip architecture. Andreou places particular emphasis on software-architecture-hardware co-design.

A theory for architecture exploration, given a set of instructions with specified parallelism and energy-delay cost, is also shown. Flexible architectures, to meet specialized needs, can be implemented using chiplets, with 4 nano-Abacus chiplets designed. Several examples of event-based processing and other techniques needed to achieve a Bio-Inspired System Architecture for Energy Efficient (BIGDATA) Computing with application to wide area motion imagery are reviewed.

1 AI in "The Atlantic" (era 2012-2018)

NOVEMBER 2018 ISSUE

The interesting questions are not about the stupidity of AI and what not!
They are the questions about social, regulatory and ethical issues and WE the makers of Intelligent Machines must have something to say!

How the Enlightenment Ends - The Atlantic
Jun 15, 2018 ... Philosophically, intellectually—in every way—human society is unprepared for the rise of artificial intelligence.

Can Artificial Intelligence Be Smarter Than a Person? -
Sep 28, 2018 ... The Spooky Genius of Artificial Intelligence. AI doesn't ... The framial robot AILA [artificial intelligence lightweight android] operates a a ...

Artificial Intelligence Shows Why Atheism is Unpopular -
Jul 23, 2016 ... It's a noble idea: If leaders can use artificial intelligence to predict which policy will produce the best outcomes, maybe we'll end up with a ...

The Dawn of the Age of Artificial Intelligence - The
Feb 14, 2014 ... We're going to see artificial intelligence do more and more, and as this happens costs will go down, outcomes will improve, and our lives will ...

Why Self-Taught Artificial Intelligence Has Trouble With
Feb 21, 2018 ... The latest artificial intelligence systems start from zero knowledge of a game and grow to world-beating in a matter of hours. But researchers ...

Artificial Intelligence Opens the Vatican Secret Archives
Apr 30, 2018 ... Known as In Codice Ratio, it uses a combination of artificial intelligence and optical-character-recognition (OCR) software to scour these ...

'Artificial Intelligence' Has Become Meaningless - The
Mar 4, 2017 ... In science fiction, the premise or threat of artificial intelligence is tied to humans' relationship to conscious machines. Whether it's Terminators or ...

China's Rise in Artificial Intelligence - The Atlantic
Feb 16, 2017 ... A woman is silhouetted against the Baidu logo Chinese tech company Baidu has invested heavily in artificial intelligence research. Aly Song / ...

Noam Chomsky: Where Artificial Intelligence Went
Nov 1, 2012 ... Noam Chomsky on Where Artificial Intelligence Went Wrong. An extended conversation with the legendary linguist. Yarden Katz. Nov 1, 2012.

When PARRY Met ELIZA: A Ridiculous Chatbot
Jun 8, 2014 ... Whether this marks the first beating of the Turing Test, the pioneering computer scientist's trial for artificial intelligence, remains a matter of ...

The End of Reality - The Atlantic
May 15, 2018 ... Everything south of the neck, however, belongs to different women. An artificial intelligence has almost seamlessly stitched the familiar visages ...

The Future of AI Depends on High-School Girls - The
May 23, 2018 ... Women make up one-quarter of computer scientists. But in the field of artificial intelligence those numbers are likely much lower.

Reconstructing Work
Pessimists believe that humans will be made redundant by artificial intelligence [AI] and robots, leaving them unable to find work in a future bereft of jobs.

All Stories by Henry A. Kissinger - The Atlantic
Jun 15, 2018 ... Philosophically, intellectually—in every way—human society is unprepared for the rise of artificial intelligence. Henry A. Kissinger - June 2018 ...

Siri, Cortana, Alexa: Why Do So Many Digital Assistants
Mar 30, 2016 ... ... in which apps and web services will be replaced by artificial intelligence. "As we start to see these intelligent agents," Mortensen told me, "the ...

Automation Disproportionately Affects Latinos' Jobs -
Dec 21, 2017 ... Who will the biggest victims be in this new age of automation, in which artificial intelligence dominates and tasks such as driving are ...

Review: AMC's TV Show 'Humans' Could Be the Ultimate
Jun 26, 2015 ... ... from both fiction and the real world, mankind itself finds the question of "why" a silly one when it comes to the notion of artificial intelligence.

---

2 Artificial Intelligence

Merriam Webster

Intelligence:
   (1) the ability to learn or understand or to deal with new or trying situations
   (2) the ability to apply knowledge to manipulate one's environment or to think abstractly as measured by objective criteria (such as tests)
   (3) the ability to perform computer functions

Synonyms: acumen, astuteness, caginess (also cageyness), canniness, clear-sightedness, foxiness, shrewdness, wit, hardheadedness, keenness, knowingness, sharpness

Andreas G. Andreou
http:/andreoulab.net/
@andreoulab

Artificial:
   (1) humanly contrived
   (2) lacking in natural or spontaneous quality
   (3) based on differential morphological characters not necessarily indicative of natural relationships

Synonyms: affected, assumed, bogus, contrived, factitious, fake, false, feigned, forced, mechanical, mock, phony, plastic, pretended, pseudo, put-on, sham, simulated, strained, unnatural

3   **Three Waves of A.I. (I)**

<u>First Wave:</u> Handcrafted Knowledge (rule based systems),
Enables reasoning over limited domains and poor handling of
uncertainty, no learning.
Chess playing, IBM deep blue chess playing computer
LISP, Macsyma, C, Pico LISP
LISP machines, Symbolics, LMI Lambda, Connection Machine,
PilMCU

<u>Second Wave:</u> Statistical learning of low dimensional
structures from high dimensional data, Deep Neural
Networks, Deep Learning
Game of Go, AlpaGo, Google deep mind
TensorFlow, Caffe, MSFT Cognitive Toolkit, Theano, Intel
Deep Learning Toolkit, ...
GPUs, TPU, ...

ARTICLE   doi:10.1038/nature16961

**Mastering the game of Go with deep
neural networks and tree search**

David Silver[1]*, Aja Huang[1]*, Chris J. Maddison[1], Arthur Guez[1], Laurent Sifre[1], George van den Driessche[1], Julian Schrittwieser[1], Ioannis Antonoglou[1], Veda Panneershelvam[1], Marc Lanctot[1], Sander Dieleman[1], Dominik Grewe[1], John Nham[2], Nal Kalchbrenner[1], Ilya Sutskever[2], Timothy Lillicrap[1], Madeleine Leach[1], Koray Kavukcuoglu[1], Thore Graepel[1] & Demis Hassabis[1]

4   **Three Waves of A.I. (II)**

<u>Third Wave:</u> Cognitive Computing, model based systems with
contextual adaptation and composability; models to drive and
explain decisions
Generative models (for example Bayesian)

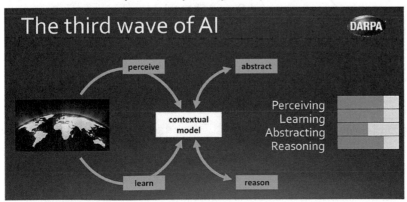

A DARPA perspective on
Artificial Intelligence

John Lunchbury
Director I2O, DARPA

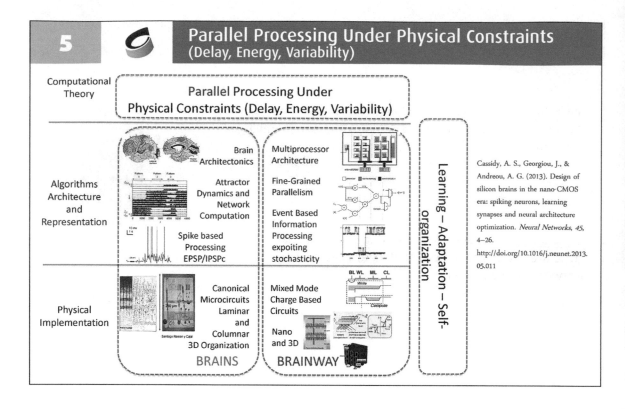

5 — Parallel Processing Under Physical Constraints (Delay, Energy, Variability)

Computational Theory: Parallel Processing Under Physical Constraints (Delay, Energy, Variability)

Algorithms Architecture and Representation

BRAINS: Brain Architectonics; Attractor Dynamics and Network Computation; Spike based Processing EPSP/IPSPc

BRAINWAY: Multiprocessor Architecture; Fine-Grained Parallelism; Event Based Information Processing expoiting stochasticity

Physical Implementation

BRAINS: Canonical Microcircuits Laminar and Columnar 3D Organization

BRAINWAY: Mixed Mode Charge Based Circuits; Nano and 3D

Learning – Adaptation – Self-organization

Cassidy, A. S., Georgiou, J., & Andreou, A. G. (2013). Design of silicon brains in the nano-CMOS era: spiking neurons, learning synapses and neural architecture optimization. *Neural Networks*, *45*, 4–26. http://doi.org/10.1016/j.neunet.2013.05.011

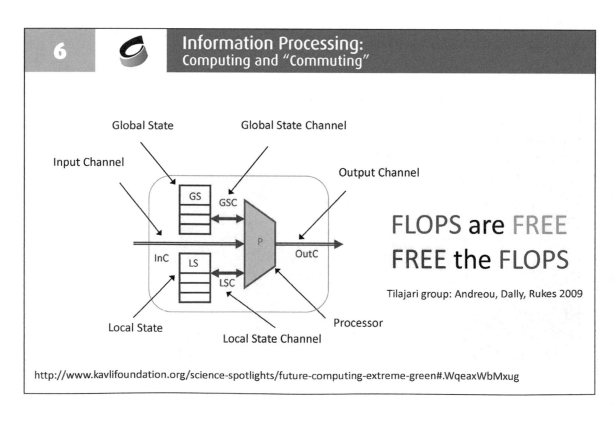

6 — Information Processing: Computing and "Commuting"

Global State
Global State Channel
Input Channel
Output Channel
GS
GSC
InC
LS
LSC
P
OutC
Local State
Local State Channel
Processor

FLOPS are FREE
FREE the FLOPS

Tilajari group: Andreou, Dally, Rukes 2009

http://www.kavlifoundation.org/science-spotlights/future-computing-extreme-green#.WqeaxWbMxug

7 — Why 3D?

Tezzaron TSVs

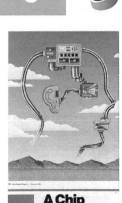

IBM 14nm FinFET Embedded DRAM

Crossbar ReRAM

Everspin Embedded MRAM

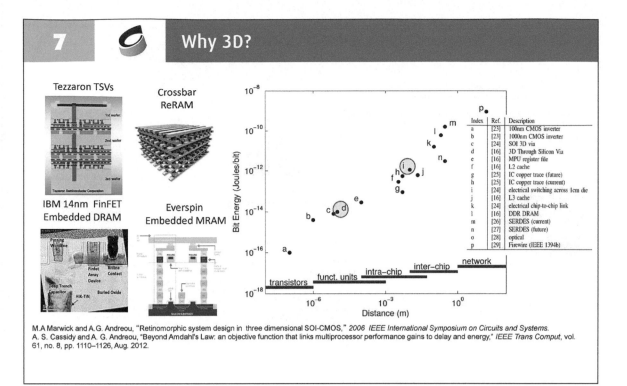

Index	Ref.	Description
a	[23]	100nm CMOS inverter
b	[23]	1000nm CMOS inverter
c	[24]	SOI 3D via
d	[16]	3D Through Silicon Via
e	[16]	MPU register file
f	[16]	L2 cache
g	[25]	IC copper trace (future)
h	[25]	IC copper trace (current)
i	[24]	electrical switching across 1cm die
j	[16]	L3 cache
k	[24]	electrical chip-to-chip link
l	[16]	DDR DRAM
m	[26]	SERDES (current)
n	[27]	SERDES (future)
o	[28]	optical
p	[29]	Firewire (IEEE 1394b)

M.A Marwick and A.G. Andreou, "Retinomorphic system design in three dimensional SOI-CMOS," *2006 IEEE International Symposium on Circuits and Systems.*
A. S. Cassidy and A. G. Andreou, "Beyond Amdahl's Law: an objective function that links multiprocessor performance gains to delay and energy," *IEEE Trans Comput*, vol. 61, no. 8, pp. 1110–1126, Aug. 2012.

8 — Early 90s

A Chip You Can Talk To
... and it will obey, right away.
That is the goal.
Oh, and it ought to be small enough to fit in your phone.

Marc Cohen (left) has a chip that can pick out one voice from a din. Philippa Pouliquen's chip will link the others: it's an associative memory.

technology," says Andreou. He thinks the major obstacles lie in the fact that scientists have only a general understanding of how the brain works. "We only know," he says, "what happens at the very early stages of the sensory pathways, such as the cochlea, or the retina... Besides, who said we were trying to build a brain?"

What they're trying to build is analog microchips that can perform spe-

For computers to become useful tools for the masses, they must be redesigned, so they communicate not via the keyboard, but directly with their operators and their surroundings. They must obey the spoken word, recognize faces and objects, even read handwriting, Andreou reasons. To do that, they need to be given sight and hearing, possibly also touch and smell.

Kwabena A. Boahen is working toward the B.S./M.S.E. degree in electrical engineering at the Johns Hopkins University, Baltimore, MD. His current research interests are Analog VLSI design and testing, with applications in synthetic neural and sensory systems. Mr. Boahen is a member of Tau Beta Pi.

Compute environment:

Pentium Pro, 0.5um, 5.5 M xtors
150 MHz, FSB: 66MHz, 3.3 Volts on package 256KB L2 cache, 3.3V 35W, 64 MB RAM, 5GB HD

DEC Alpha 21064A, 0.5um, , 2.85 M xtors 200 MHz, 256KB Bcache, 3.3 V, 30W, 64MB RAM, 5GB HD

CAD environment: Magic, IRSIM, SPICE

Nowak, R., & Andreou, A. G., A chip you can talk to, *Johns Hopkins Magazine*, 30–38, December 1990.

9 Embedded Analog Computing in Digital Memories

Winner-Takes-All Associative Memory: A Hamming Distance Vector Quantizer

PHILIPPE O. POULIQUEN,[1] ANDREAS G. ANDREOU[1], AND KIM STROHBEHN[2]

Analog Integrated Circuits and Signal Processing, 13, 211–222 (1997)

$$\max_{i=1,N}\left(\frac{|\vec{x} \wedge \vec{c}_i|}{a+|\vec{c}_i|}\right)$$

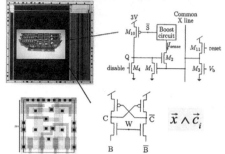

$\vec{x} \wedge \vec{c}_i$

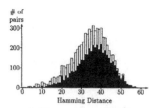

exploiting problem statistics!

pose processor.) In an DEC-Alpha based general purpose computer it takes 10000 cycles to do a single pattern matching computation and thus it takes a total of $20\mu s$ per classification. Power dissipation is 30W at 500 MHz and therefore the energy per classification is 600μJ. The Pentium-Pro is worse, because it requires 30W at 150 MHz and more than 10000 cycles for a single pattern matching. In contrast, the total current in the WAM is: $(124 \times 116 \times 10)$ nA continuous bias current for the memory cells at 5V. Computation time is approximately $70\mu s$ for a total energy per classification of approximately 100 nJ. The power dissipation in

1. Memory and processing are integrated in a single structure; this is analogous to the synapse in biology.
2. The system has an internal model that is related to the problem to be solved (*prior* knowledge). This is the template set of patterns to be classified.
3. The system is capable of learning i.e. templates can be changed to adapt to a different character set (different problem). This is done at the expense of storage capacity—we use a RAM based cell instead of a more compact ROM cell–.
4. The system processes information in a parallel and hierarchical fashion in a variable precision architecture. I.e. given the statistics of the problem, most of the computation is carried out with low precision (three or four bit) analog hardware. Yet arbitrary precision computation is possible through recursive processing that exploits a programmable WTA (capability to mask specific bits in the winner takes all circuitry).
5. The system is fault tolerant and gracefully degrades. The same structures that is used in the *precision-on-demand* architecture can also be used to reconfigure the system for defects in the fabrication process. The components of the chip that are worse matched can be disabled during operation.

2um CMOS technology, 5Volts

10 Cognitive Computing Technology in Data Centers

Each Tier: "Infinitely" Scalable, Heterogeneous

Datacenter / Cloud Tier

Mainframes, Unix Servers | Google WSC Servers *

Aggregation Tiers

Edge Devices

Inter- and Intra-Tier Networking

Multiple Tiers, from Edge to Datacenter

Internet, digital telephony, email, and other electronic media are making human communication, activity, and experience to be mediated by computers. Future projected needs in data centers are data intensive applications in Cognitive Computing Technology (CCT).

CCT aims at advancing intelligent software and hardware that can process, analyze, and distill knowledge from vast quantities of text, speech, images and biological data ultimately with and as much nuance and depth of understanding as a human would.

The Datacenter as a Computer, 2nd Edition, by L. Barrosa et al., 2013

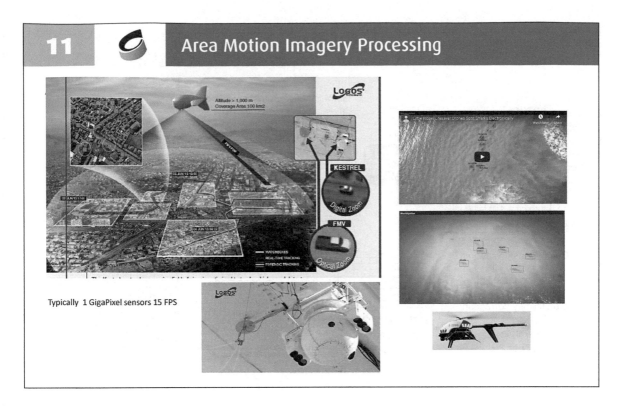

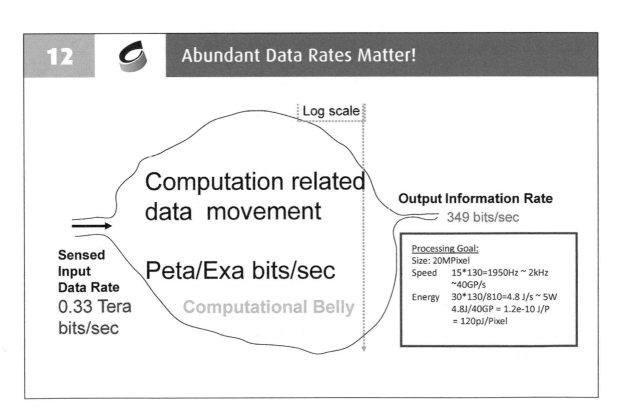

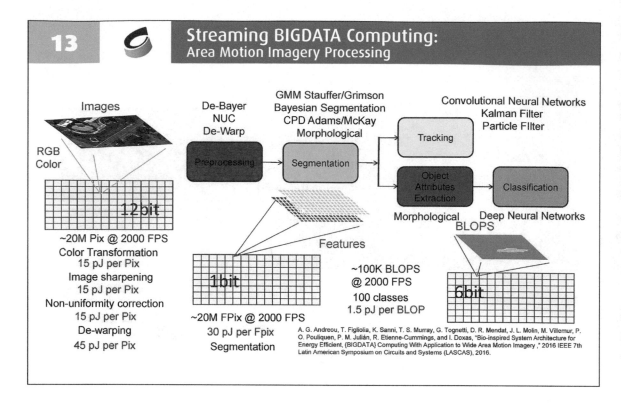

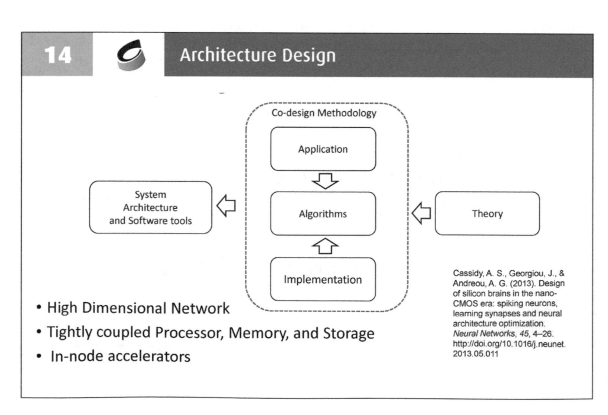

15 — Computing at the Energy Limit

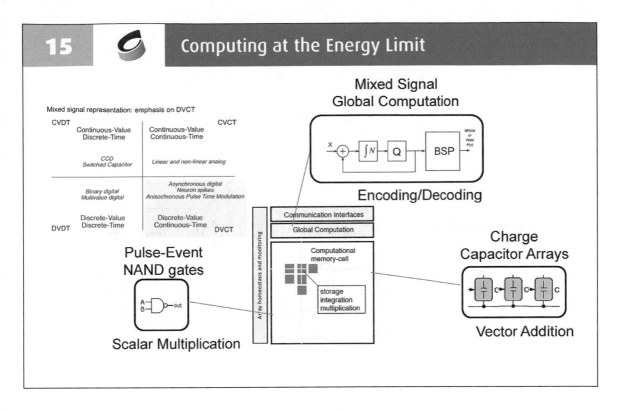

16 — Design Methodology

- Precision on demand: accuracy of computations can be configured before runtime by setting the number of events for every circuit and that defines the precision.
- Digital hardware that can be operated at **Ultra Low Voltage at the limit ~400mV**
- Analog hardware that can be operated at **Ultra Low Voltage at the limit ~500mV**
- The data could represent probabilities directly if desired (if results are normalized to one).
- Mixed-signal circuits for probability encoding and decoding
- Seamless integration with industry standard design flows (Cadence, Synopsys, Mentor Graphics)
- Design in cost-effective CMOS node (55nm GF) using industry standard IP (i.e ARM, SST SuperFlash ESF3) but with our own CMOS library cells, SRAM and I/O pads.

17   **Architecture Design**

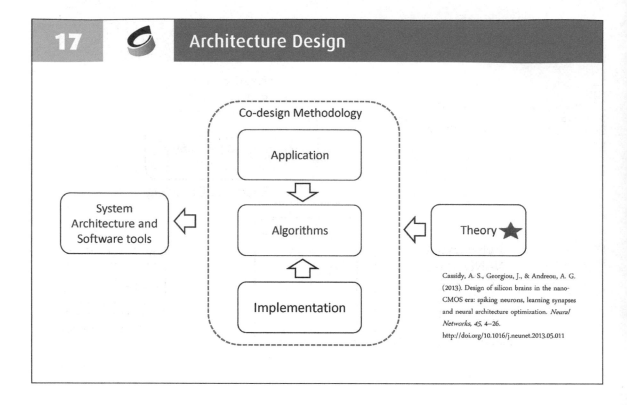

Co-design Methodology

Application

System Architecture and Software tools

Algorithms

Theory ★

Implementation

Cassidy, A. S., Georgiou, J., & Andreou, A. G. (2013). Design of silicon brains in the nano-CMOS era: spiking neurons, learning synapses and neural architecture optimization. *Neural Networks, 45,* 4–26. http://doi.org/10.1016/j.neunet.2013.05.011

18   **A Theory for Architecture Exploration. Why?**

Because computer architectures today are mostly done through a process of deep omphaloskepsis

IEEE JOURNAL OF SOLID-STATE CIRCUITS

KiloCore: A 32-nm 1000-Processor Computational Array

Brent Bohnenstiehl, *Student Member, IEEE*, Aaron Stillmaker, *Member, IEEE*, Jon J. Pimentel, *Student Member, IEEE*, Timothy Andreas, *Student Member, IEEE*, Bin Liu, *Student Member, IEEE*, Anh T. Tran, *Member, IEEE*, Emmanuel Adeagbo, *Student Member, IEEE*, and Bevan M. Baas, *Senior Member, IEEE*

19 A Theory for Architecture Exploration

$$\text{minimize} \quad J = \sum_{j=0}^{K-1} F_j N_j^{\gamma-1} \sum_{i=0}^{M-1} G_{ij} D_i E_{ij}^{\gamma}$$

Example: Chip Multi Processors; single chip area fixed and in 2D or 3D stack (fine or course grain)

$$A_{tot} \quad = \quad N(A_P + A_{L1} + A_{L2} + A_{IC}) + A_{I/O} + A_{memIF}$$

- F_j – fraction of instructions with j^{th} parallelism
- N_j – parallelism degree of jth fraction
- G_i – fraction of instructions with i^{th} energy-delay cost
- D_i – i^{th} delay cost
- E_i – i^{th} energy cost

Cassidy, A. S., & Andreou, A. G. (2012). Beyond Amdahl's Law: an objective function that links multiprocessor performance gains to delay and energy. *IEEE Transactions on Computers, 61*(8), 1110–1126. http://doi.org/10.1109/TC.2011.169

20 Architecture Exploration Flow

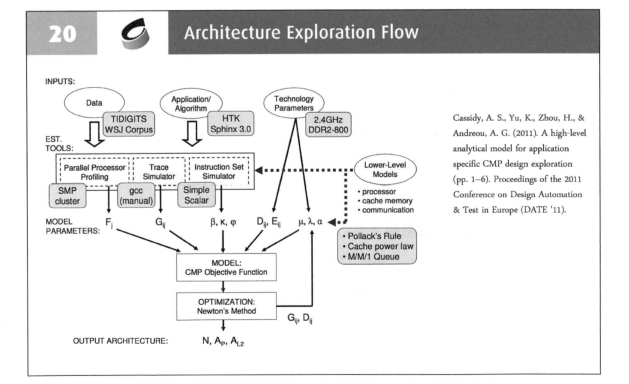

Cassidy, A. S., Yu, K., Zhou, H., & Andreou, A. G. (2011). A high-level analytical model for application specific CMP design exploration (pp. 1–6). Proceedings of the 2011 Conference on Design Automation & Test in Europe (DATE '11).

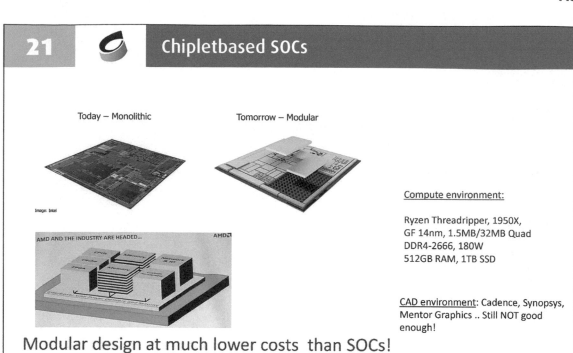

21 — Chipletbased SOCs

Today – Monolithic

Tomorrow – Modular

Image: Intel

AMD AND THE INDUSTRY ARE HEADED...

AMD

Compute environment:

Ryzen Threadripper, 1950X,
GF 14nm, 1.5MB/32MB Quad
DDR4-2666, 180W
512GB RAM, 1TB SSD

CAD environment: Cadence, Synopsys,
Mentor Graphics .. Still NOT good
enough!

Modular design at much lower costs than SOCs!

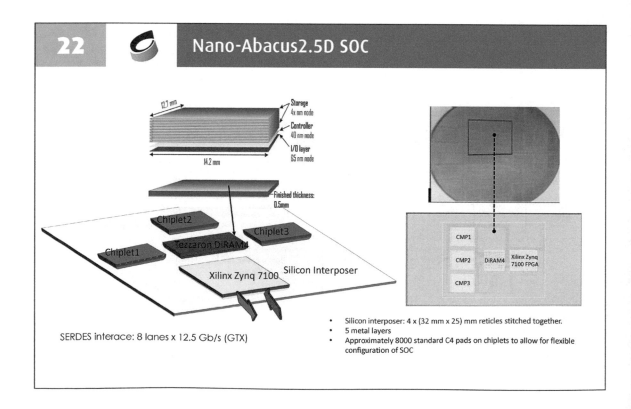

22 — Nano-Abacus2.5D SOC

12.7 mm

Storage
4x nm node

Controller
40 nm node

I/O layer
65 nm node

14.2 mm

Finished thickness:
0.5mm

Chiplet2

Chiplet3

Chiplet1

Tezzaron DiRAM4

Xilinx Zynq 7100   Silicon Interposer

SERDES interace: 8 lanes x 12.5 Gb/s (GTX)

CMP1

CMP2   DiRAM4   Xilinx Zynq
7100 FPGA

CMP3

- Silicon interposer: 4 x (32 mm x 25) mm reticles stitched together.
- 5 metal layers
- Approximately 8000 standard C4 pads on chiplets to allow for flexible
configuration of SOC

23 Chiplet Architecture (16 X 7 Cores)

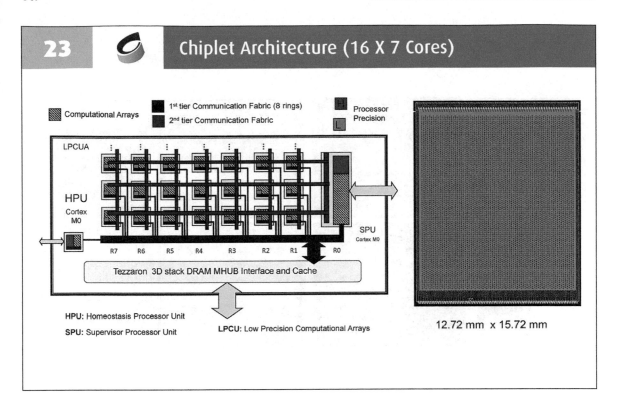

24 Nano-Abacus Processing Units and Software Stack

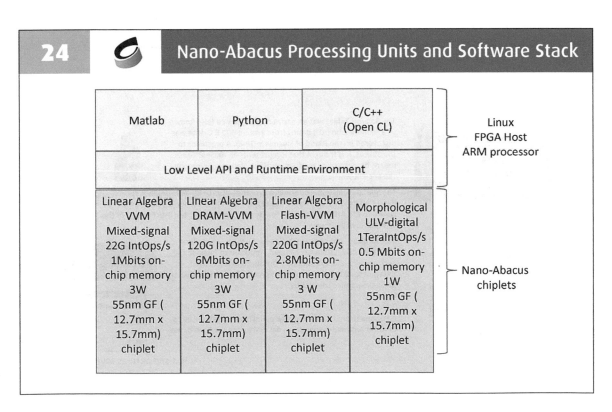

25 The Four Nano-Abacus Chiplets

26 Calculating/Computing by Counting!

1/100000

The genesis of
nanoAbacus CMP

The **Salamis Tablet** was an early counting device (also known as a "counting board") dating from around 300 B.C. that was discovered on the island of Salamis in 1846. A precursor to the abacus, it is thought that it represents an Ancient Greek means of performing mathematical calculations common in the ancient world.

The **Suanpan**, also spelled souanpan is an abacus of Chinese origin first described in a 190 CE book of the Eastern Han Dynasty, namely Supplementary Notes on the Art of Figures written by Xu Yue. However, the exact design of this suanpan is not known.

The **Soroban** is an abacus developed in Japan. It is derived from the ancient Chinese suanpan, imported to Japan in the 14th century. Like the suanpan, the soroban is still used today, despite the proliferation of practical and affordable pocket electronic calculators.

A **Yupana** (from Quechua yupay: count) is an abacus used to perform arithmetic operations dating back to the time of the Incas.

Text and pictures source Wikipedia

27 — Streaming Bigdata Computing:
Area Motion Imagery Processing

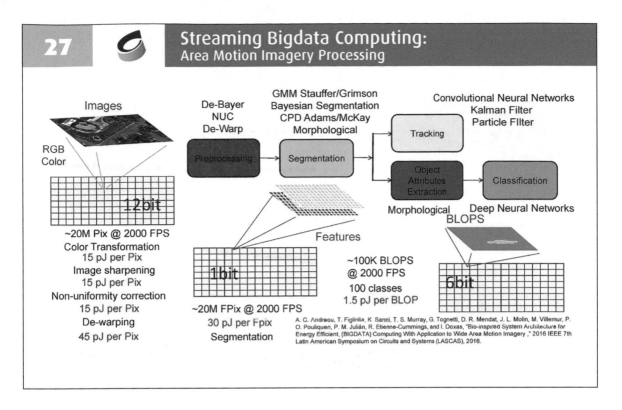

Images

RGB Color

12bit

~20M Pix @ 2000 FPS
Color Transformation
15 pJ per Pix
Image sharpening
15 pJ per Pix
Non-uniformity correction
15 pJ per Pix
De-warping
45 pJ per Pix

De-Bayer
NUC
De-Warp

GMM Stauffer/Grimson
Bayesian Segmentation
CPD Adams/McKay
Morphological

Preprocessing → Segmentation

1bit

~20M FPix @ 2000 FPS
30 pJ per Fpix
Segmentation

Features

~100K BLOPS
@ 2000 FPS
100 classes
1.5 pJ per BLOP

Convolutional Neural Networks
Kalman Filter
Particle Filter

Tracking

Object Attributes Extraction → Classification

Morphological   Deep Neural Networks
BLOPS

6bit

A. G. Andreou, T. Figlinlia, K. Sanni, T. S. Murray, G. Tognetti, D. R. Mendat, J. L. Molin, M. Villemur, P. O. Pouliquen, P. M. Julián, R. Etienne-Cummings, and I. Doxas, "Bio-inspired System Architecture for Energy Efficient, [BIGDATA] Computing With Application to Wide Area Motion Imagery ," 2016 IEEE 7th Latin American Symposium on Circuits and Systems (LASCAS), 2016.

28 — Event-based Small (3X3) Kernel Convolutions (De-Bayer)

De-Bayering a Gigapixel image

Image De-Bayer and Color Transform with VVM5 Chip
64 samples (6-bit stochastic precision)

Approximately 4-bit output precision

Image De-Bayer and Color Transform with VVM5 Chip
256 samples (8-bit stochastic precision)

Approximately 5-bit output precision

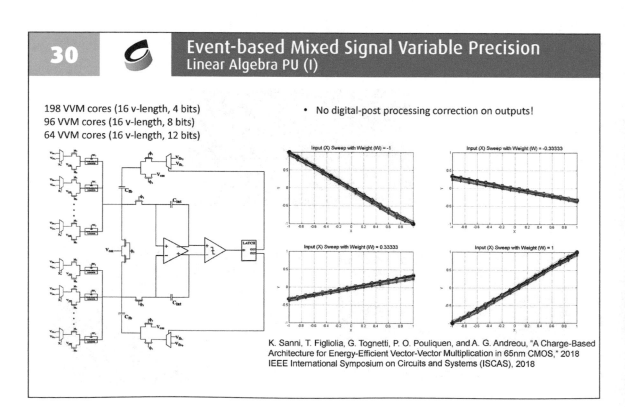

29 — Event-based Mixed Signal Convolution Processing

K. Sanni, T. Figliolia, G. Tognetti, P. O. Pouliquen, and A. G. Andreou, "A Charge-Based Architecture for Energy-Efficient Vector-Vector Multiplication in 65nm CMOS," presented at the ISCAS 2018, 2018

30 — Event-based Mixed Signal Variable Precision Linear Algebra PU (I)

198 VVM cores (16 v-length, 4 bits)
96 VVM cores (16 v-length, 8 bits)
64 VVM cores (16 v-length, 12 bits)

- No digital-post processing correction on outputs!

K. Sanni, T. Figliolia, G. Tognetti, P. O. Pouliquen, and A. G. Andreou, "A Charge-Based Architecture for Energy-Efficient Vector-Vector Multiplication in 65nm CMOS," 2018 IEEE International Symposium on Circuits and Systems (ISCAS), 2018

31   Event-based Mixed Signal Variable Precision
Linear Algebra PU (II)

A Mixed-Signal Successive Approximation Architecture for Energy-Efficient Fixed-Point Arithmetic in 16nm FinFET, International Conference Circuits and Systems (ISCAS 2019)

SAR Logic

Simulated Characteristic of the SAR Mulitply-Add Single Core

Conv. Mult-Add
SAR Mult-Add

Energy/Op (J)

Supply Voltage (V)

Process	16nm FinFET
Operation	8-bit Multiply-Add
Supply Voltage	0.4V
Clock Frequency	50MHz
Throughput	3.57MOPs
Power	24.5nW
Energy/Op	6.85fJ
Energy Efficiency	146TOPs/W

32   Streaming BIGDATA Computing:
Area Motion Imagery Processing

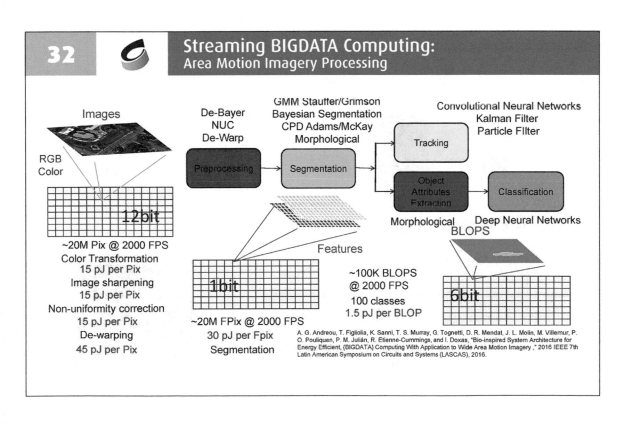

Images

RGB
Color

12bit

~20M Pix @ 2000 FPS
Color Transformation
15 pJ per Pix
Image sharpening
15 pJ per Pix
Non-uniformity correction
15 pJ per Pix
De-warping
45 pJ per Pix

De-Bayer
NUC
De-Warp

Preprocessing

1bit

~20M FPix @ 2000 FPS
30 pJ per Fpix
Segmentation

GMM Stauffer/Grimson
Bayesian Segmentation
CPD Adams/McKay
Morphological

Segmentation

Features

Convolutional Neural Networks
Kalman Filter
Particle FIlter

Tracking

Object
Attributes
Extraction

Classification

Morphological   Deep Neural Networks
BLOPS

~100K BLOPS
@ 2000 FPS
100 classes
1.5 pJ per BLOP

6bit

A. G. Andreou, T. Figliolia, K. Sanni, T. S. Murray, G. Tognetti, D. R. Mendat, J. L. Molin, M. Villemur, P. O. Pouliquen, P. M. Julián, R. Etienne-Cummings, and I. Doxas, "Bio-inspired System Architecture for Energy Efficient, (BIGDATA) Computing With Application to Wide Area Motion Imagery," 2016 IEEE 7th Latin American Symposium on Circuits and Systems (LASCAS), 2016.

33 — Exact Bayesian Inference and Online Learning

Gaussian process where mu takes two values with probability given by Bernoulli RV theta

Ryan. P. Adams and D. J. MacKay, "Bayesian Online Changepoint Detection," *arXiv.org*, vol. stat.ML. 19-Oct-2007.

Change Variable (RV)

Markov Process

Data

When posterior distributions $p(\theta|D)$ are in the same family as the prior probability distribution $p(\theta)$, then prior and posterior are called conjugate distributions; the prior is called a **conjugate prior** for the likelihood function

$$P(\Theta / D) = \frac{P(D / \Theta)P(\Theta)}{P(D)}$$

34 — Event-Based Exact Bayesian Inference Computation for Background Foreground Detection (Online Change Point Detection)

3072 (64 x 48) Bayesian processing cores

$I(x,y,t=1)$ $I(x,y,t=N)$

Pixel state to be computed (background or foreground

$I(x,y)=\{I(x,y,t)\}_{t=1,...N}$
Observed pixel value at frame t

$$p(r_t, x_{1:t})$$

t=1 $t=1,...,N$ Frames   t=N

Signal with traditional CPD

CPD chip

~ 10 nJ per Bayesian inference –CPD op-

35 Exploiting Stochasticity at All Levels

Probabilistic Algorithms
- Exact Bayesian (Change Point Detection)

Probabilistic Computational Structures
- Stochastic architecture
- Probabilistic event based representation
- Probabilistic mixed-signal circuits

Probabilistic nano-CMOS
- Random Telegraph Noise

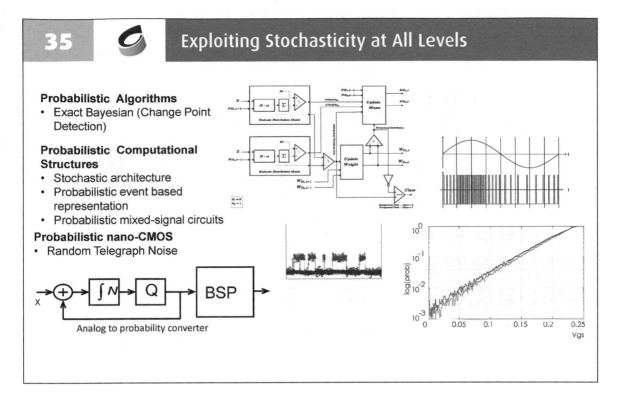

Analog to probability converter

36 Streaming BIGDATA Computing:
Area Motion Imagery Processing

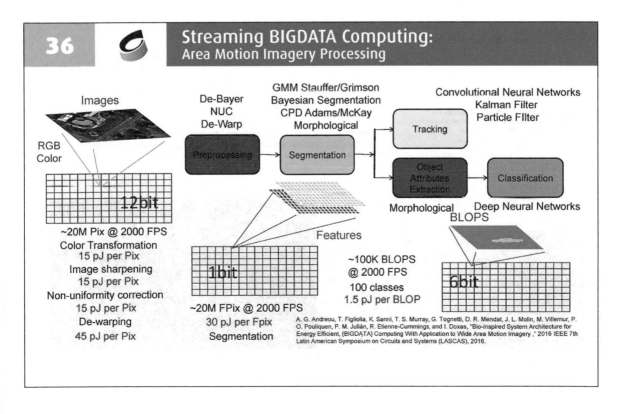

Images

RGB Color

12bit

~20M Pix @ 2000 FPS
Color Transformation
15 pJ per Pix
Image sharpening
15 pJ per Pix
Non-uniformity correction
15 pJ per Pix
De-warping
45 pJ per Pix

De-Bayer
NUC
De-Warp

Preprocessing

1bit

~20M FPix @ 2000 FPS
30 pJ per Fpix
Segmentation

GMM Stauffer/Grimson
Bayesian Segmentation
CPD Adams/McKay
Morphological

Segmentation

Features

~100K BLOPS
@ 2000 FPS
100 classes
1.5 pJ per BLOP

Convolutional Neural Networks
Kalman Filter
Particle Filter

Tracking

Object
Attributes
Extraction

Classification

Morphological   Deep Neural Networks
BLOPS

6bit

A. G. Andreou, T. Figliolia, K. Sanni, T. S. Murray, G. Tognetti, D. R. Mendat, J. L. Molin, M. Villemur, P. O. Pouliquen, P. M. Julián, R. Etienne-Cummings, and I. Doxas, "Bio-inspired System Architecture for Energy Efficient, {BIGDATA} Computing With Application to Wide Area Motion Imagery ," 2016 IEEE 7th Latin American Symposium on Circuits and Systems (LASCAS), 2016.

37 — Segmentation–morphological Processing (I)

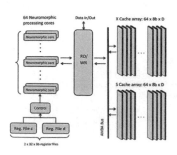

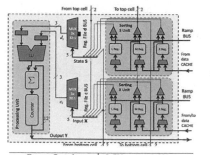

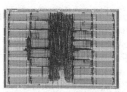

Technology	55nm	
Core Size	$0.76mm \times 1.13mm$	
Transistors	1.57 million	
Array size	64×64	

M. Villemur, P. Julian, A. G. Andreou, and T. Figliolia, Neuromorphic Cellular Neural Network Processor for Intelligent Internet-of-Things, 2018 IEEE International Symposium on Circuits and Systems (ISCAS), 2018.

Power Supply	0.5V	1.2V
Clock Frequency	1MHz	12MHz
Power (dynamic)	$1.88\mu W$	$3.18\mu W$
Power (leakage)	$5.50\mu W$	$33.80\mu W$
Power (total)	$7.38\mu W$	$36.98\mu W$
MOPS	32.8	393
E/OP [fJ]	225	94.2
E/OP [fJ] (dynamic)	57.4	8.1
TOPS/W	4.43	10.6
TOPS/W (dynamic)	17.4	123

1024 x 16 x 7 = 114688 PU
~ 1 TOPS at 100mW

38 — Segmentation–morphological Processing (II)

Energy aware simplicial processor for embedded morphological visual processing in intelligent internet of things

M. Villemur, P. Julian[✉] and A. G. Andreou

This Letter presents the architecture implementation and testing of an single instruction multiple data (SIMD) processor for energy aware embedded morphological visual processing using the simplicial piece-wise linear approximation. The architecture comprises a linear array of 48 × 48 processing elements, each connected to an eight-neighbour clique operating on binary input and state data. The architecture is synthesised from a custom designed ultra low-voltage CMOS library and fabricated in a 55 nm CMOS technology. The chip is capable of dynamic voltage/frequency scaling with power supplies between 0.5 and 1.2 V. The fabricated chip achieves an overall performance of 293 TOPS/W with dynamic energy dissipation efficiency of 3.4 fJ per output operation at 0.6 V.

M. Villemur, P. Julian, and A. G. Andreou, "Energy aware simplicial processor for embedded morphological visual processing in intelligent internet of things," *IET Electronics Letters*, vol. 54, no. 7, pp. 420–422, Apr. 2018.

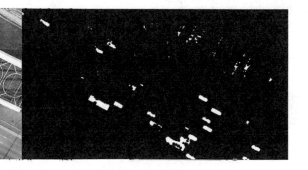

39 Machine Learning FLASH – SST ESF3

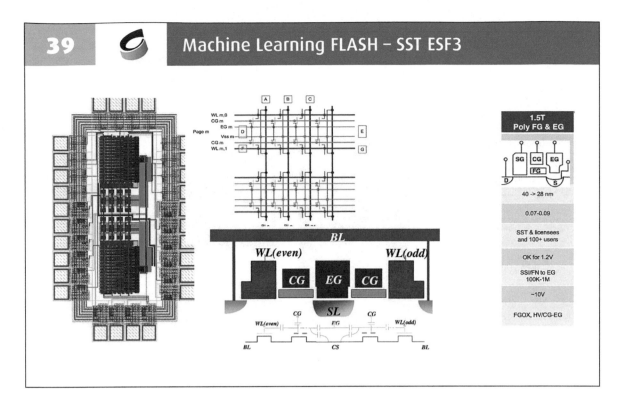

40 Analog and Multibit Programming
of Digital Flash Cells

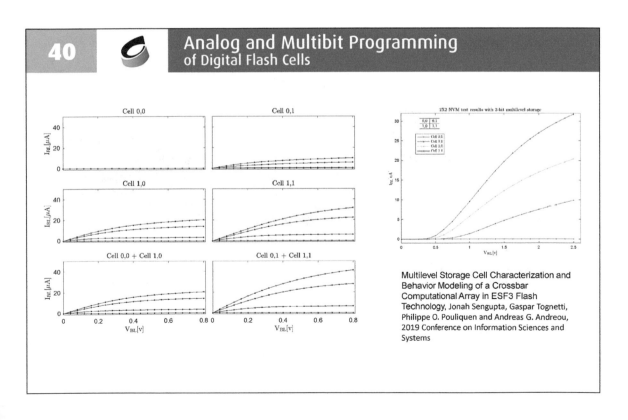

Multilevel Storage Cell Characterization and Behavior Modeling of a Crossbar Computational Array in ESF3 Flash Technology, Jonah Sengupta, Gaspar Tognetti, Philippe O. Pouliquen and Andreas G. Andreou, 2019 Conference on Information Sciences and Systems

41 PCIe blades 2 x FMC Boards Per Card

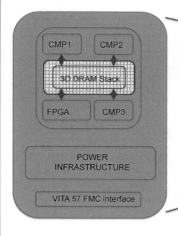

CMP1 CMP2

3D DRAM Stack

FPGA CMP3

POWER INFRASTRUCTURE

VITA 57 FMC interface

nanoAbacus
HPC-2.5D SOC

JOHNS HOPKINS

nanoAbacus
HPC-SOC

42 NeuromorphicSpike Based Symbolic Computation

Graph Analytics on
the SpiNNaker
System

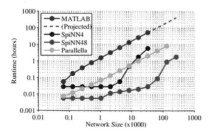

- MATLAB
- (Projected)
- SpiNN4
- SpiNN48
- Parallella

Runtime (hours)

Network Size (x1000)

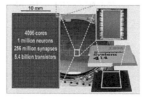

10 mm

4096 cores
1 million neurons
256 million synapses
5.4 billion transistors

Systems
414

Real-time Scalable Cortical Computing at
46 Giga-Synaptic OPS/Watt with
~100× Speedup in Time-to-Solution and
~100,000× Reduction in Energy-to-Solution

SIMILARITY RESULTS FOR FOUR QUERIES ON TRUENORTH
WITH 95,000 WORD DICTIONARY.

telluride	basketball	mountains	neuron
carbon	volleyball	landforms	electron
colorado	handball	ranges	protein
copper	soccer	glacier	tissue
springs	ncaa	plateau	cells

D. R. Mendat, S. Chin, S. B. Furber, and A. G. Andreou,
"Neuromorphic Sampling on the SpiNNaker and Parallella Chip
Multiprocessors," Proceedings of the 2016 IEEE 7th Latin American
Symposium on Circuits and Systems (LASCAS), 2016, pp. 399–402.

Word2vec Word Similarities on IBM's TrueNorth Neurosynaptic System , Daniel R. Mendat∗†,
Andrew S. Cassidy‡, Guido Zarrella§, and Andreas G. Andreou∗† BioCAS 2018

43 Our People

www.andreoulab.net   Twitter: @andreoulab   connect with us

44 Our Sponsors

Mapping the Cardiac Acousteome: Biosensing and Computational Modeling Applied to Smart Diagnosis and Monitoring of Heart Conditions NSF-SCH INT-20132017

Signals to Symbols: From Bio-inspired Hardware to Cognitive Systems, NSF-INSPIRE: SMA-1248056

SIDEARM: Sensor Inference and Decision Engine via Adaptive Regulation of Memory arrays with BAE systems, DARPA-MTO UPSIDE HR0011-13-C-0051

PerSEUS: Peta-op SupErcomputing Unconventional System DARPA-DSO seedling with UCLA, Dartmouth and NG (JHU-prime)

COMPROIC: Computational ROIC Tradeoffs DARPA-MTO seedling

 DECISIVE: DROIC with Event Based Cognitive-Processing of Imagery and Spikes Integrated Vertically with NG, DARPA-MTO ReImagine (kickoff meeting June 2017)

Multichannel Vestibular Prosthesis Pilot Early, Feasibility Trial, NIH-NIDCD R01DC13536 –Charley Delasantina PI-

CHAPTER 08

The Loihi Neuromorphic Research Chip

Mike Davies

Intel Labs

In this chapter Mike Davies gives an industry perspective on neuromorphic computing, indicating that the research is not limited to universities, but is also being pursued by industrial research organizations. The primary focus of the chapter is Intel's Loihi neuromorphic chip. Future directions in neuromorphic research are also covered.

Although Loihi is a digital chip, this avenue of research pushes beyond conventional von Neumann architectures. Davies describes Loihi as an exploratory AI research chip inspired by the principles of neural computation in nature. Loihi explores a very different paradigm compared to the von Neumann and matrix-acceleration architectures in wide use today. The Loihi architecture integrates self-modifying processes allowing autonomous adaptation and learning in response to discrete events and an evolving statistical environment. The chip supports sparse and irregular communication between its units in the form of packetized event-driven "spikes". Loihi is implemented using a novel asynchronous design methodology that allows fully exploiting activation sparsity.

Exciting recent results suggest that neuromorphic chips that combine a fine-grain, spike-based parallel neural network architecture and asynchronous design can provide compelling gains in computing efficiency, scalable to orders of magnitude, for the right kinds of adaptive dynamic problems.

1 — Neuromorphic Computing Exploration Space

"Deep Learning" /
Artificial Neural Networks

Conventional Machine Learning

Brain-Inspired Computation

Competitive Computer Architectures

Novel neuro-inspired computation

Research Goals:

- **Broad class** of brain-inspired computation
- **Efficient** hardware implementations
- **Scalable** from small to large problems and systems

Examples:

- Online and lifelong learning
- Learning without cloud assistance
- Learning with sparse supervision
- Understanding spatiotemporal data
- Probabilistic inference and learning
- Sparse coding/optimization
- Nonlinear adaptive control (robotics)
- Pattern matching with high occlusion
- SLAM and path planning
- Dynamical systems modeling

2 — Architectural Landscape

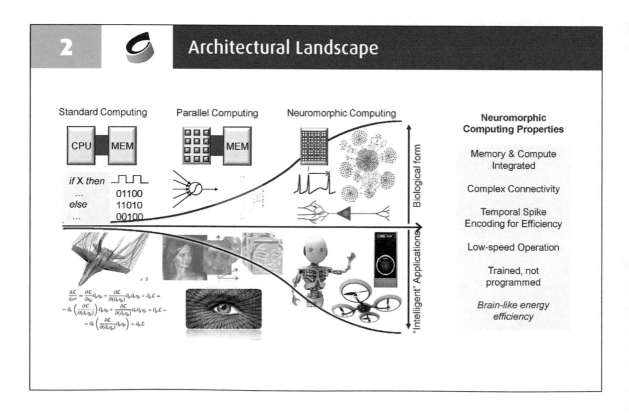

Standard Computing | Parallel Computing | Neuromorphic Computing

Neuromorphic Computing Properties

Memory & Compute Integrated

Complex Connectivity

Temporal Spike Encoding for Efficiency

Low-speed Operation

Trained, not programmed

Brain-like energy efficiency

3 The Engineering Perspective

- Nature has come up with something amazing. Let's copy it...

- Not so simple – very different design regimes

- Yet objectives and constraints are largely the same...
    - Energy minimization
    - Fast response time
    - Cheap to produce

Need to understand and apply the basic principles, *adapting for differences*

Status today:

	Nature	Silicon	Ratio
Neuron density[1]	100k/mm²	5k/mm²	20x
Synaptic area[1]	0.001 um²	0.4 um²[2]	400x
Synaptic Op Energy	~2 fJ	~4 pJ	2000x

But... [1] Planar neocortex [2] ~5b SRAM

	Nature	Silicon	Ratio
Max firing rate	100 Hz	1 GHz	10,000,000x
Synaptic error rate	75%	0%	∞

Nature	Silicon
Autonomous self-assembly	Fabricated manufacturing
Per-instance variability desired	Variability causes brittle failures
Limited plasticity over lifetime	Must support rapid reprogramming
Nondeterministic operation	Deterministic operation desired

4 Some Principles of Neural Computation

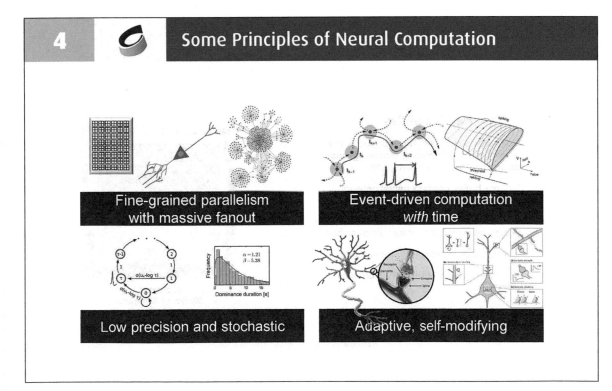

Fine-grained parallelism with massive fanout

Event-driven computation *with* time

Low precision and stochastic

Adaptive, self-modifying

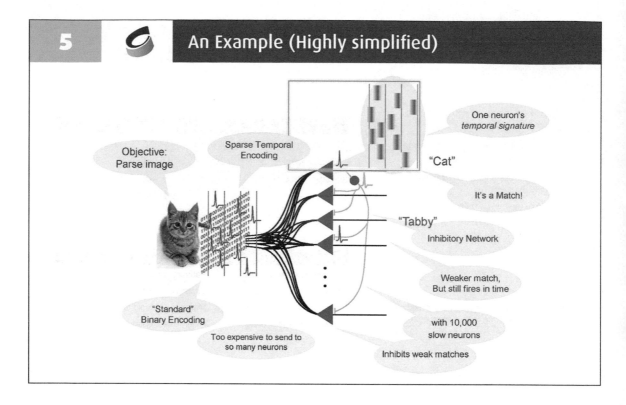

5 — An Example (Highly simplified)

Objective: Parse image

Sparse Temporal Encoding

One neuron's *temporal signature*

"Cat"

It's a Match!

"Tabby"

Inhibitory Network

Weaker match, But still fires in time

"Standard" Binary Encoding

Too expensive to send to so many neurons

with 10,000 slow neurons

Inhibits weak matches

6 — Our Loihi research chip

Key Properties

- 128 neuromorphic cores supporting up to 128k neurons and 128M synapses with an **advanced spiking neural network feature set.**

- Supports **highly complex neural network topologies**

- **Scalable on-chip learning** capabilities to support an unprecedented range of learning algorithms

- Fully digital **asynchronous** implementation

- Fabricated in Intel's **14nm FinFET process** technology

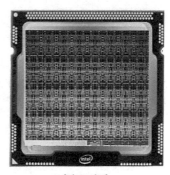

**Integrated
Memory + Compute
Neuromorphic Architecture**

M Davies et al., "Loihi: A Neuromorphic Manycore Processor with On-Chip Learning", IEEE MICRO, Jan/Feb 2018.

7   Learning with Synaptic Plasticity

- **Local learning rules** – essential property for efficient scalability

- Rules derived by **optimizing an emergent statistical objective**

- Plasticity on **wide range of time scales** for
    - ✓ Immediate supervised (labelled) learning
    - ✓ Unsupervised self-organization
    - ✓ Working memory
    - ✓ Reinforcement-based delayed feedback

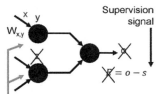

Learning rules for weight $W_{x,y}$ may *only* access presynaptic state x and postsynaptic state y

However *reward spikes* may be used to distribute graded reward/punishment values to a particular set of axon fanouts

8   Loihi's Trace-Based Programmable Learning

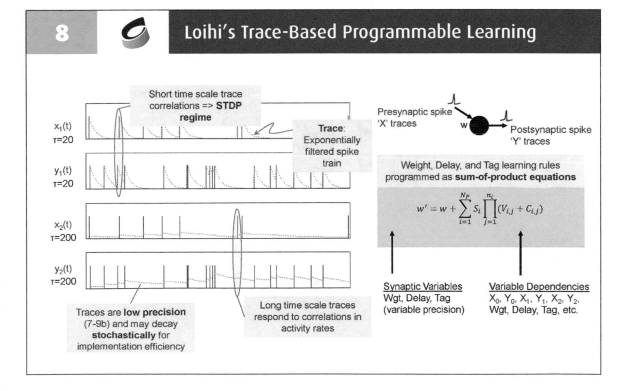

9 Case Study: LASSO Sparse Coding

Problem

$$\min_{z} \frac{1}{2} \|x - Dz\|_2^2 + \lambda \|z\|_1$$

Input — Reconstruction — Sparse regularization

Implementation

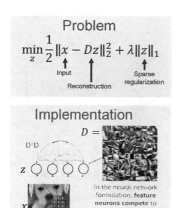

$D =$

D^TD

z

x

In the neural network formulation, **feature neurons compete** to reconstruct image with as few contributors as possible

Tang et al, arxiv: 1705:05475

LASSO Optimization Using the *Spiking Locally Competitive Algorithm*

Inhibition

$$-\left(\boldsymbol{d}_i^T \cdot \boldsymbol{d}_j\right) z_j$$

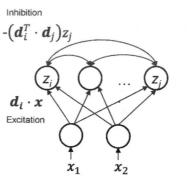

z_i ... z_j

$\boldsymbol{d}_i \cdot \boldsymbol{x}$

Excitation

x_1 x_2

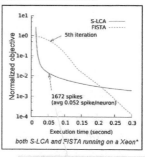

1672 spikes
(avg 0.052 spike/neuron)

both S-LCA and FISTA running on a Xeon*

Neuromorphic algorithm rapidly finds a near-optimal solution

* Performance results are based on testing as of May 2017 and may not reflect all publicly available security updates. No product can be absolutely secure.

---

10 Spiking LCA Dynamics on a Loihi Predecessor

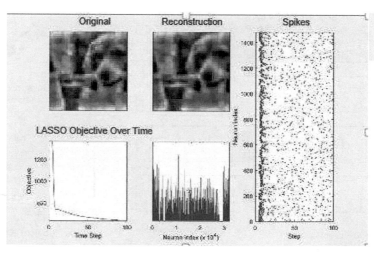

Intense but very brief period of competition

Much faster convergence on a neuromorphic architecture

Original · Reconstruction · Spikes

LASSO Objective Over Time

11   Sparse Coding Results: N1 vs Atom CPU

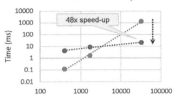

Time to Solution Comparison

48x speed-up

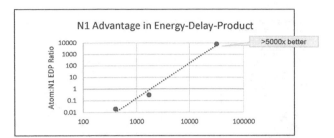

N1 Advantage in Energy-Delay-Product

>5000x better

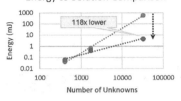

Energy to Solution Comparison

118x lower

Number of Unknowns

Comparison of sparse coding on N1 versus the FISTA* LASSO solver on an Atom CPU**

* Best conventional LASSO solver (LARS also evaluated)   • Atom (FISTA)
** Iso-process, roughly iso-area (6-10mm²)   • N1
   PTPX-based measurements

Performance results are based on testing as of May 2017 and may not reflect all publicly available security updates. No product can be absolutely secure.

12   Our "Hello World" Application

Supervised Learning for Object Recognition

Video available online: https://youtu.be/cDKnt9ldXv0

13   Demonstrates Real-time Learning on 10's of mW

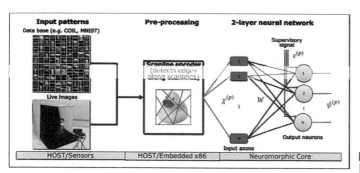

Performance on COIL20 data set

99.6% accuracy in 78 seconds

87% accuracy in 4 seconds

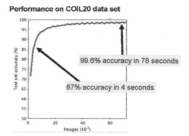

Resource Utilization	Count	Utilization
Neurons	20	0.02%
Synapses	38400	0.28%
SNN Cores	1	0.78%

Uses **less than 1%** of chip resources

	Training	Inference
Active energy per image (total)	553 uJ	128 uJ
Neuromorphic energy	322 uJ	13 uJ
Processing time per image	7.5 ms	1.8 ms
Chip power		
Neuromorphic power		

14   Adaptive Control of a Robot Arm Using Loihi

SNN adaptive dynamic controller implemented on Loihi that allows the robot arm to adjust in real time to nonlinear, unpredictable changes in system mechanics[1].

Result outperforms standard PD & PID control algorithms.

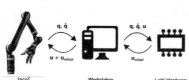

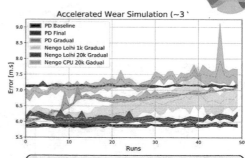

Accelerated Wear Simulation (~3 `

Different control methods adapting to a gradual, linear increase in friction, over the course of 50 runs. This simulates ~3 years of wear over the course of 16.67 minutes of run time, a 90K times speed up. Only 20K neurons on Loihi is able to successfully cope with this perturbation.

[1] DeWolf, T., Stewart, T. C., Slotine, J. J., & Eliasmith, C. (2016, November). A spiking neural model of adaptive arm control. In *Proc. R. Soc. B* (Vol. 283, No. 1843, p. 20162134). The Royal Society.

15 Other Novel Algorithms Supported by Loihi

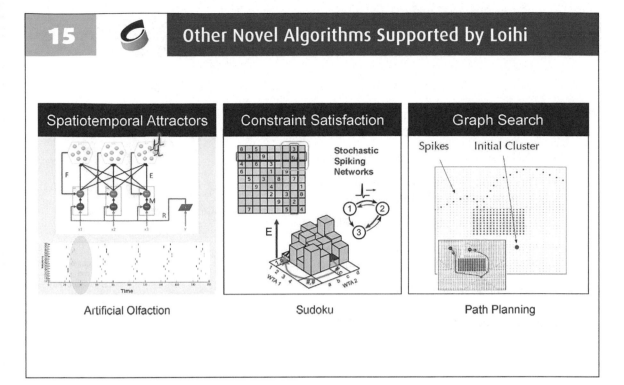

Spatiotemporal Attractors	Constraint Satisfaction	Graph Search
Artificial Olfaction	Sudoku	Path Planning

16 Graph Search – Path Planning

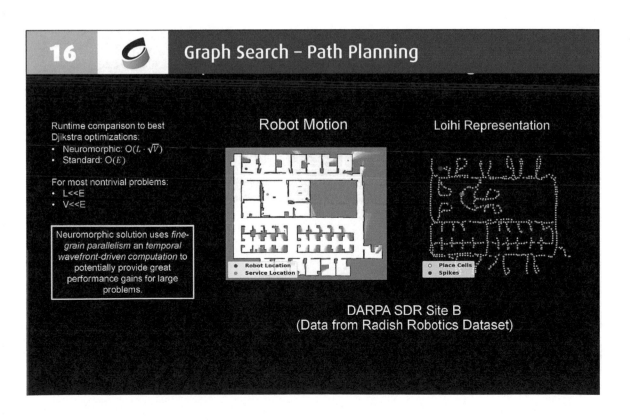

Runtime comparison to best Djikstra optimizations:
- Neuromorphic: $O(L \cdot \sqrt{V})$
- Standard: $O(E)$

For most nontrivial problems:
- $L \ll E$
- $V \ll E$

Neuromorphic solution uses *fine-grain parallelism* an *temporal wavefront-driven computation* to potentially provide great performance gains for large problems.

Robot Motion

Loihi Representation

DARPA SDR Site B
(Data from Radish Robotics Dataset)

17 — Spike-based LSTMs – "LSNNs"

Simple adaptive spiking model achieves LSTM-level accuracy

- SNN reservoir augmented with adaptive neurons

- Thresholds rise on each spike, decay exponentially
  ☞ *Highly energy-efficient adaptation*

- Trained offline with BPTT (TensorFlow)

- Achieves 96% accuracy on sequential MNIST, same as equivalent LSTMs

- Now being ported to Loihi

[Bellec et al, arXiv preprint arXiv:1803.09574]

Performance comparison

First case of an
**SNN matching
LSTM accuracy**

18 — Our Future Outlook: Engaging the Community

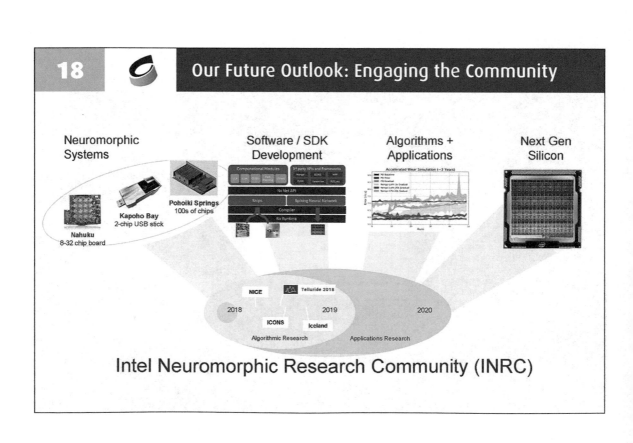

Neuromorphic Systems

Nahuku
8-32 chip board

Kapoho Bay
2-chip USB stick

Pohoiki Springs
100s of chips

Software / SDK Development

Algorithms + Applications

Next Gen Silicon

NICE   Telluride 2018

2018   2019   2020

ICONS   Iceland

Algorithmic Research   Applications Research

Intel Neuromorphic Research Community (INRC)

19   Mike Davies, Intel

Depends on timeframe – AI is not really "AI" let alone "smart AI" today

Depends on definition of "smart"

- <u>In reach today</u>: Phenomenal pattern matching & generation
- <u>Smart behavior</u>: Quick learning, able to create new knowledge and articulate complex reasoning in convincing ways

Backpropagation + SGD

- Excellent offline training algorithm, not the key to smart AI

Need new algorithms

- <u>Some in sight</u>: HD computing (VSAs), olfaction-inspired one-shot learning, learning-to-learn with adaptive spiking neural nets
- Need neuromorphic architectures to efficiently scale & run these

20   Legal Information

This presentation contains the general insights and opinions of Intel Corporation ("Intel"). The information in this presentation is provided for information only and is not to be relied upon for any other purpose than educational. Intel makes no representations or warranties regarding the accuracy or completeness of the information in this presentation Intel accepts no duty to update this presentation based on more current information. Intel is not liable for any damages, direct or indirect, consequential or otherwise, that may arise, directly or indirectly, from the use or misuse of the information in this presentation.

Intel technologies' features and benefits depend on system configuration and may require enabled hardware, software or service activation. Learn more at intel.com, or from the OEM or retailer.

No computer system can be absolutely secure. No license (express or implied, by estoppel or otherwise) to any intellectual property rights is granted by this document. Intel, the Intel logo and Xeon are trademarks of Intel Corporation in the United States and other countries.

*Other names and brands may be claimed as the property of others.

Copyright © 2018 Intel Corporation.

CHAPTER 09

RRAM Fabric for Neuromorphic Computing Applications

Wei Lu

University of Michigan

As suggested in earlier chapters, in-memory computing using a crossbar circuit is a promising approach for addressing AI compute challenges. In this chapter Wei Lu describes recent research progress on RRAM (resistive random access memory) devices and chip-level design and fabrication. Lu shows how neural networks can be mapped into RRAM crossbar arrays, providing a platform for AI computing. The chapter presents a number of chips for AI computing tasks fabricated using an integrated RRAM/CMOS technology. Lu also presents new research directions, including, "bio-realistic" hardware using ionic dynamics to emulate physical ionic/molecular processes in biology as well as a general-purpose programmable in-memory processing hardware fabric.

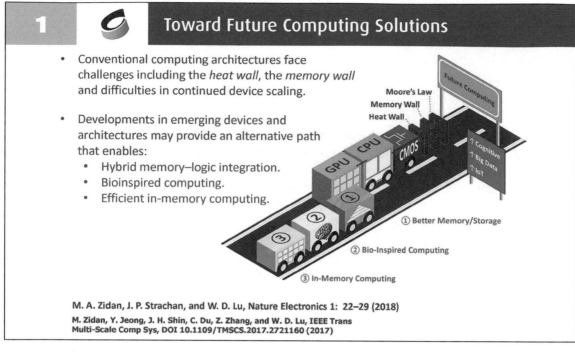

1 — Toward Future Computing Solutions

- Conventional computing architectures face challenges including the *heat wall*, the *memory wall* and difficulties in continued device scaling.

- Developments in emerging devices and architectures may provide an alternative path that enables:
  - Hybrid memory–logic integration.
  - Bioinspired computing.
  - Efficient in-memory computing.

Moore's Law
Memory Wall
Heat Wall

GPU CPU CMOS

① Cognitive
① Big Data
① IoT

① Better Memory/Storage

② Bio-Inspired Computing

③ In-Memory Computing

M. A. Zidan, J. P. Strachan, and W. D. Lu, Nature Electronics 1: 22–29 (2018)

M. Zidan, Y. Jeong, J. H. Shin, C. Du, Z. Zhang, and W. D. Lu, IEEE Trans Multi-Scale Comp Sys, DOI 10.1109/TMSCS.2017.2721160 (2017)

The development of future computing circuits and systems will require close collaboration between device researchers and computer architects and circuit designs, to maximally utilize the unique properties of emerging memory and logic devices and mask device nonidealities.

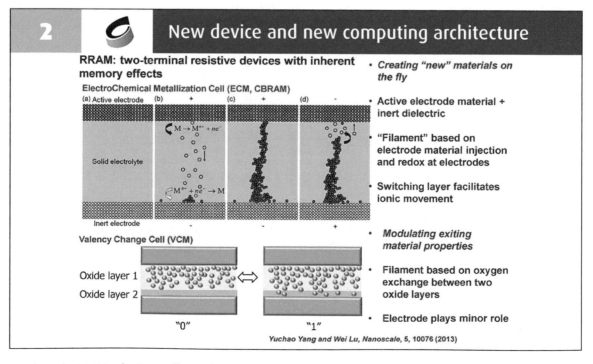

2 — New device and new computing architecture

RRAM: two-terminal resistive devices with inherent memory effects

ElectroChemical Metallization Cell (ECM, CBRAM)

(a) Active electrode   (b) +   (c) +   (d) −

$M \rightarrow M^{n+} + ne^-$

Solid electrolyte

$M^{n+} + ne^- \rightarrow M$

Inert electrode

Valency Change Cell (VCM)

Oxide layer 1

Oxide layer 2

"0"   "1"

- *Creating "new" materials on the fly*

- **Active electrode material + inert dielectric**

- **"Filament" based on electrode material injection and redox at electrodes**

- **Switching layer facilitates ionic movement**

- *Modulating exiting material properties*

- **Filament based on oxygen exchange between two oxide layers**

- **Electrode plays minor role**

Yuchao Yang and Wei Lu, Nanoscale, 5, 10076 (2013)

Ion-based RRAM devices allow the materials' properties to be modulated on-the-fly, using simply a voltage bias in a circuit. The change in material properties, e.g. resistance, in turn allows these devices to be used as promising memory and even logic devices.

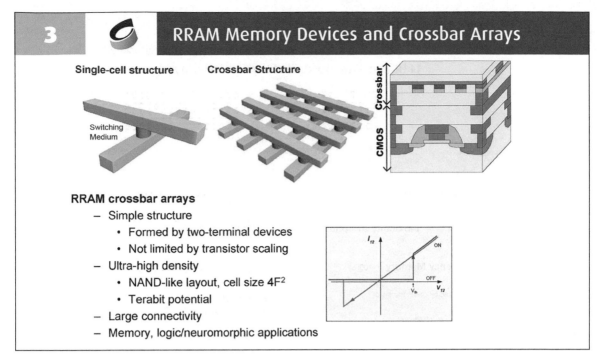

The simple two-terminal structure allows RRAM devices to be integrated in the very high density crossbar form. Since RRAM devices are not based on crystalline Si substrates, they can be integrated over existing logic circuitry to further improve the storage and compute density.

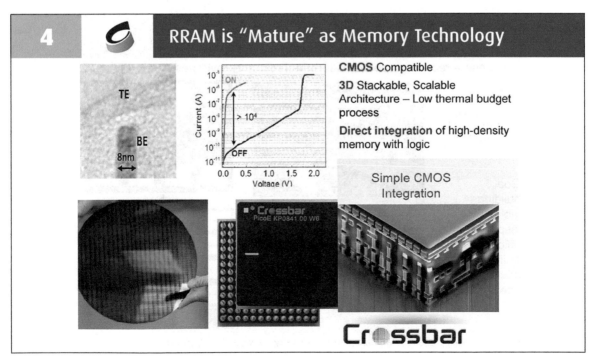

RRAM is at the verge of large scale commercialization, and is being offered by startups such as Crossbar Inc, and large foundries such as TSMC.

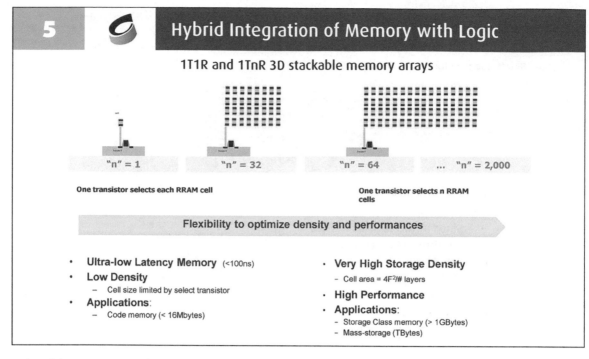

The ability to integrate high-density and fast NVM on chip increases the communication bandwidth and reduces the data path between memory and logic. This approach will most likely be employed first, by bringing memory closer to logic.

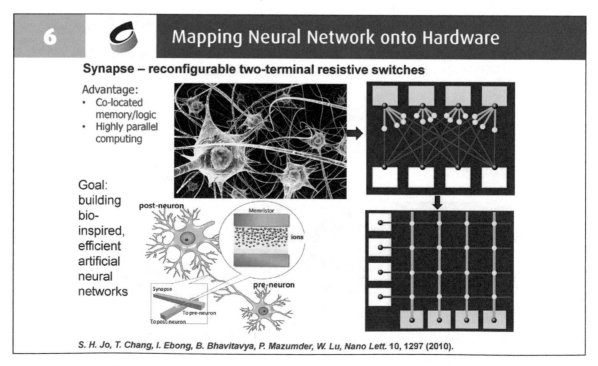

The programmable conductance in RRAM also allows the device to both store a weight and modulate the signal transmitted through it. In this regard, a single device emulates a synapse in a neural network. The neural network topology can then be readily mapped onto the crossbar structure.

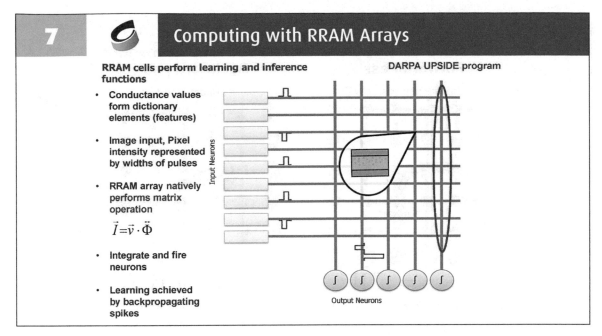

7   **Computing with RRAM Arrays**

RRAM cells perform learning and inference functions

- **Conductance values form dictionary elements (features)**

- **Image input, Pixel intensity represented by widths of pulses**

- **RRAM array natively performs matrix operation**

$$\vec{I}=\vec{v} \cdot \ddot{\Phi}$$

- **Integrate and fire neurons**

- **Learning achieved by backpropagating spikes**

DARPA UPSIDE program

Input Neurons

Output Neurons

Mathematically, the RRAM cell computes an output based on the input voltage and the stored conductance (weight), i.e. through Ohm's law. The total current collected at a column in the crossbar then equals to the dot-product of the input vector and the weight vector, based on Kirchhoff's current law. By measuring the currents at the columns in parallel, the vector-matrix multiplication can be obtained in a single read step, i.e. O(1) vs. O(N^2) in conventional implementations.

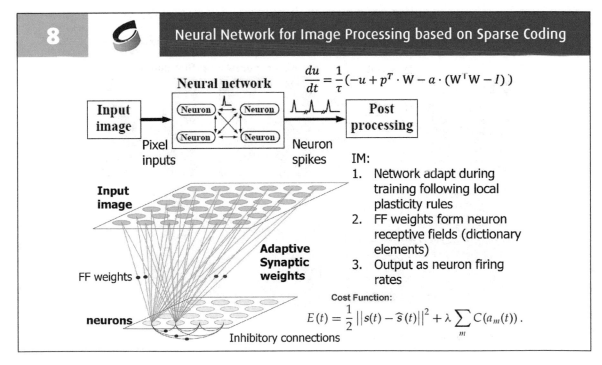

8   **Neural Network for Image Processing based on Sparse Coding**

Neural network

$$\frac{du}{dt} = \frac{1}{\tau}(-u + p^T \cdot W - a \cdot (W^T W - I))$$

Input image → Neuron ↔ Neuron / Neuron ↔ Neuron → **Post processing**

Pixel inputs   Neuron spikes

Input image

Adaptive Synaptic weights

FF weights

neurons

Inhibitory connections

IM:
1. Network adapt during training following local plasticity rules
2. FF weights form neuron receptive fields (dictionary elements)
3. Output as neuron firing rates

Cost Function:
$$E(t) = \frac{1}{2} ||s(t) - \widehat{s}(t)||^2 + \lambda \sum_m C(a_m(t)).$$

An example of image analysis through an RRAM crossbar. In this algorithm based on sparse coding, both feedforward weights and inhibition functions need to be implemented.

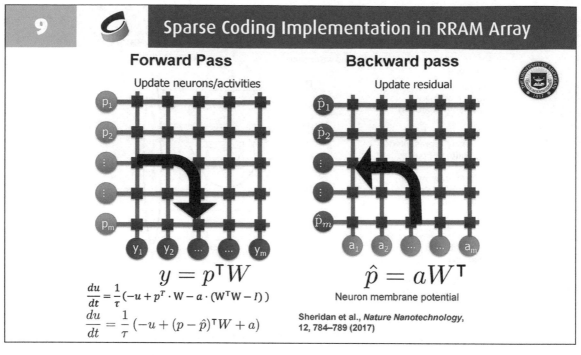

9 **Sparse Coding Implementation in RRAM Array**

Forward Pass

Update neurons/activities

$$y = p^{\mathsf{T}} W$$

$$\frac{du}{dt} = \frac{1}{\tau}(-u + p^{T} \cdot W - a \cdot (W^{T}W - I))$$

$$\frac{du}{dt} = \frac{1}{\tau}(-u + (p - \hat{p})^{\mathsf{T}}W + a)$$

Backward pass

Update residual

$$\hat{p} = aW^{\mathsf{T}}$$

Neuron membrane potential

Sheridan et al., *Nature Nanotechnology*, 12, 784–789 (2017)

The bi-directional conduction of RRAM cells allows both the conventional forward path mode and a backward path mode that produces the reconstruction based on the neuron activities and weight matrix transpose. By removing the reconstruction from the input and use the residual as the new input lateral neuron inhibition functions can be implemented.

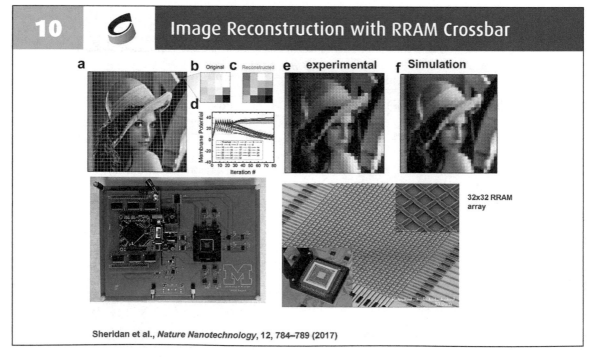

10 **Image Reconstruction with RRAM Crossbar**

a   b Original c Reconstructed e **experimental** f **Simulation**

d

Membrane Potential

Iteration #

32x32 RRAM array

Sheridan et al., *Nature Nanotechnology*, 12, 784–789 (2017)

Experimental results obtained from a 32x32 RRAM array. The neuron functions are implemented in the board.

11 — New Research Directions

- **"Biorealistic" hardware implementation**
  - using internal ionic dynamics to emulate physical ionic/molecular processes in biology
  - building circuits based on these internal dynamic processes

- **General in-memory processing hardware fabric**
  - Reprogrammable, suitable for different algorithms
  - Modular and scalable

Besides using RRAM devices as resistive elements with programmable weights, more advanced synaptic functions can be natively implemented by taking advantage of the internal RRAM ion dynamics.

Ultimately, a more general, reconfigurable computing platform can be built based on efficient RRAM in-memory computing modules.

12 — Synaptic Plasticity:
Multiple Sate Variables at Different Time Scales

NMDA receptor – activated when bound to glutamate **AND** depolarized => Ca²⁺ influx (Mg²⁺ block removed)

Ca^{2+} influx depends on the relative timing of the two processes

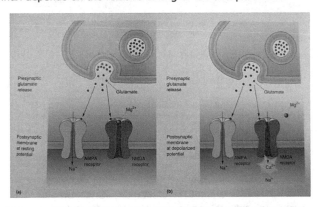

M. F. Bear, B. W. Connors, M. A. Paradiso, *Neuroscience – Exploring the Brain, 2nd ed.*, 2001

Synaptic weight changes in biology involve several processes at different time scales. The internal short-term dynamics allow the synapse to natively respond to temporal spiking patterns.

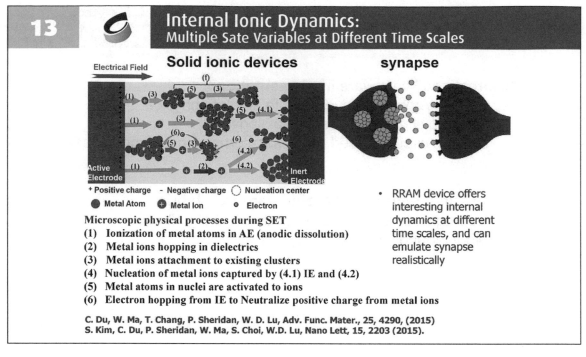

13 **Internal Ionic Dynamics:**
Multiple Sate Variables at Different Time Scales

Solid ionic devices **synapse**

Electrical Field

+ Positive charge - Negative charge ◯ Nucleation center
● Metal Atom ⊕ Metal Ion ⊖ Electron

Microscopic physical processes during SET
(1) **Ionization of metal atoms in AE (anodic dissolution)**
(2) **Metal ions hopping in dielectrics**
(3) **Metal ions attachment to existing clusters**
(4) **Nucleation of metal ions captured by (4.1) IE and (4.2)**
(5) **Metal atoms in nuclei are activated to ions**
(6) **Electron hopping from IE to Neutralize positive charge from metal ions**

- RRAM device offers interesting internal dynamics at different time scales, and can emulate synapse realistically

C. Du, W. Ma, T. Chang, P. Sheridan, W. D. Lu, Adv. Func. Mater., 25, 4290, (2015)
S. Kim, C. Du, P. Sheridan, W. Ma, S. Choi, W.D. Lu, Nano Lett, 15, 2203 (2015).

The multiple time scales within a synapse can be natively emulated in an RRAM device, by taking advantage of the different parameters involved in filament growth.

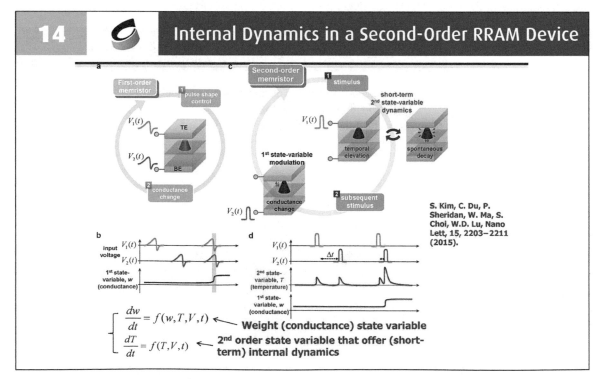

14 **Internal Dynamics in a Second-Order RRAM Device**

a First-order memristor
c Second-order memristor

$V_1(t)$ TE
$V_2(t)$ BE

pulse shape control

conductance change

stimulus

short-term 2nd state-variable dynamics

temporal elevation — spontaneous decay

1st state-variable modulation

$V_1(t)$

conductance change

subsequent stimulus

S. Kim, C. Du, P. Sheridan, W. Ma, S. Choi, W.D. Lu, Nano Lett, 15, 2203–2211 (2015).

b input voltage $V_1(t)$ $V_2(t)$
1st state-variable, w (conductance)

d $V_1(t)$ $V_2(t)$
2nd state-variable, T (temperature)
1st state-variable, w (conductance)

$$\frac{dw}{dt} = f(w,T,V,t) \leftarrow \text{Weight (conductance) state variable}$$
$$\frac{dT}{dt} = f(T,V,t) \leftarrow \text{2nd order state variable that offer (short-term) internal dynamics}$$

In a second-order RRAM, one state variable, e.g. the local temperature, exhibits short-term dynamics which in turn controls the evolution of another state variable that lead to long term weight changes.

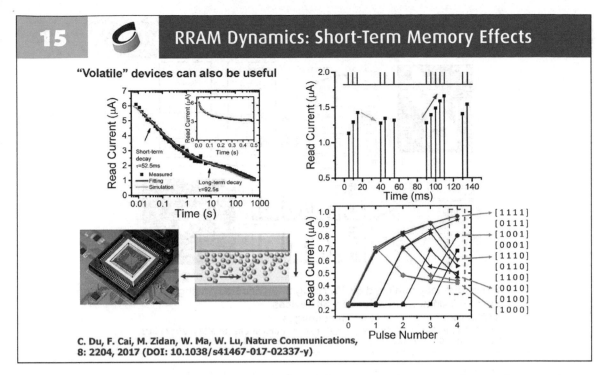

C. Du, F. Cai, M. Zidan, W. Ma, W. Lu, Nature Communications,
8: 2204, 2017 (DOI: 10.1038/s41467-017-02337-y)

Some RRAM devices only exhibit short-term memory effects. These devices can "process" temporal input data, but cannot be "trained" since the weight change is not retained.

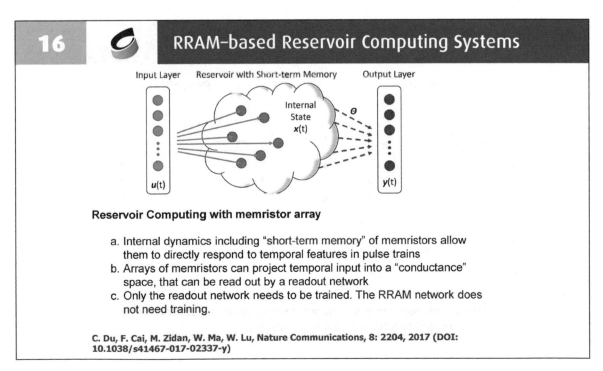

Reservoir Computing with memristor array

a. Internal dynamics including "short-term memory" of memristors allow them to directly respond to temporal features in pulse trains
b. Arrays of memristors can project temporal input into a "conductance" space, that can be read out by a readout network
c. Only the readout network needs to be trained. The RRAM network does not need training.

C. Du, F. Cai, M. Zidan, W. Ma, W. Lu, Nature Communications, 8: 2204, 2017 (DOI: 10.1038/s41467-017-02337-y)

Such devices, however, can be used in networks such as reservoir computing systems, where the reservoir layer nonlinearly maps the input but does not need to be trained. In an RC system, training only needs to be performed in the second, "readout" layer.

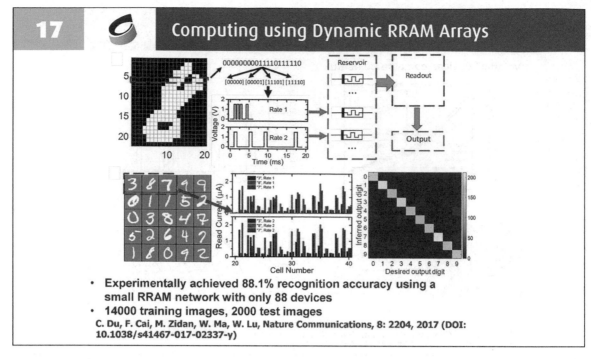

- **Experimentally achieved 88.1% recognition accuracy using a small RRAM network with only 88 devices**
- **14000 training images, 2000 test images**
  C. Du, F. Cai, M. Zidan, W. Ma, W. Lu, Nature Communications, 8: 2204, 2017 (DOI: 10.1038/s41467-017-02337-y)

Examples of RRAM-based RC systems. In this example, spatial features are converted to temporal features and fed to the RRAM based reservoir. A simple readout layer can then effectively classify the original data based on the reservoir outputs.

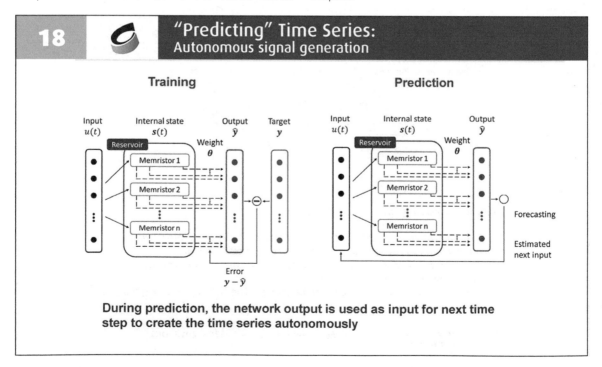

The reservoir output can also be supplied back to the network as a new input, allowing the system to reproduce/forecast the target time series autonomously.

19   New Research Directions

- **"Biorealistic" hardware implementation**
  - ○ using internal ionic dynamics to emulate physical ionic/molecular processes in biology
  - ○ building circuits based on these internal dynamic processes

- **General in-memory processing hardware fabric**
  - ○ Reprogrammable, suitable for different algorithms
  - ○ Modular and scalable

The final goal may be to build a general, reconfigurable in-memory computing platform for a broad range of data-intensive tasks.

20   New Memory, New Computing Architecture - Memory Processing Unit

General operation principle:

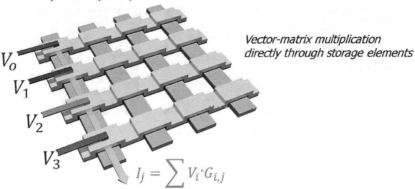

Vector-matrix multiplication directly through storage elements

$$I_j = \sum V_i \cdot G_{i,j}$$

Key concepts:
1. Computing achieved directly through the storage device, using physical principles (e.g. Ohm's law)
2. Parallel operation for the input vector and output vector
3. Both low precision and high precision computing can be obtained in the same substrate

Many data intensive computing problems can be reduced to the matrix form and solved with vector-matrix multiplications. A single RRAM-based in-memory fabric may be reconfigured for these different tasks.

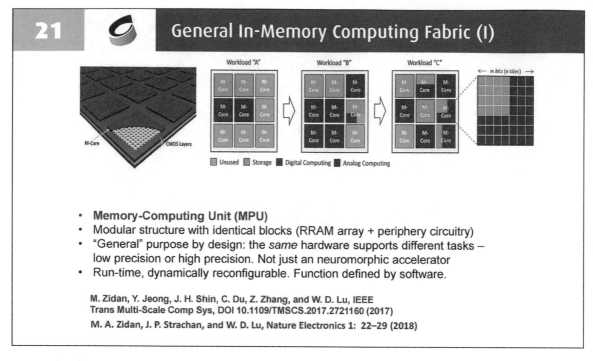

21   General In-Memory Computing Fabric (I)

- **Memory-Computing Unit (MPU)**
- Modular structure with identical blocks (RRAM array + periphery circuitry)
- "General" purpose by design: the *same* hardware supports different tasks – low precision or high precision. Not just an neuromorphic accelerator
- Run-time, dynamically reconfigurable. Function defined by software.

M. Zidan, Y. Jeong, J. H. Shin, C. Du, Z. Zhang, and W. D. Lu, IEEE
Trans Multi-Scale Comp Sys, DOI 10.1109/TMSCS.2017.2721160 (2017)
M. A. Zidan, J. P. Strachan, and W. D. Lu, Nature Electronics 1: 22–29 (2018)

Example of a general in-memory computing platform, based on an array of RRAM in-memory computing "cores".

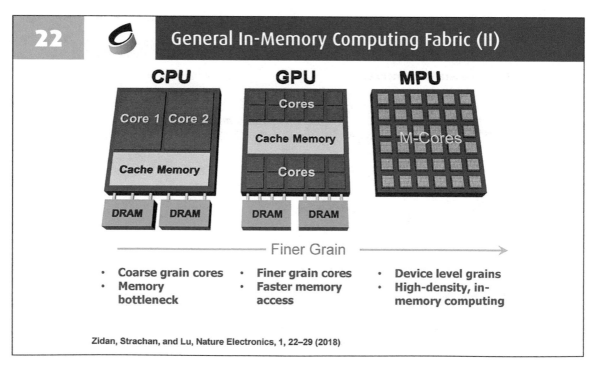

22   General In-Memory Computing Fabric (II)

Zidan, Strachan, and Lu, Nature Electronics, 1, 22–29 (2018)

This approach can be considered as an extension of the trend of moving data-intensive tasks from CPUs to GPUs. Compared with GPUs, the memory-processing unit (MPU) will offer much higher parallelism (down to the device level), and further reduced data path (i.e. in-memory compute).

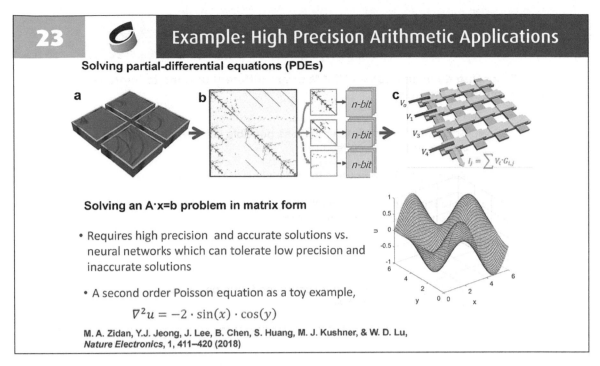

An example of using RRAM-based in-memory computing module to solve partial-differential equations, for tasks beyond neural network implementations.

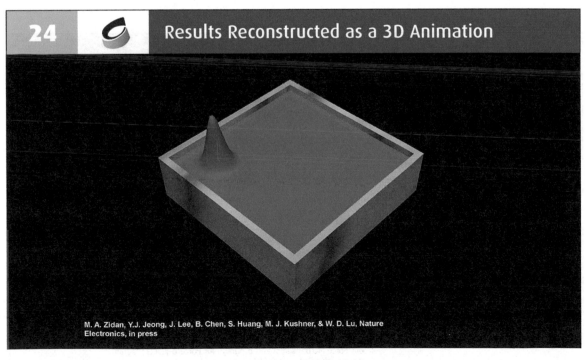

An example of a time-dependent PDE solution experimentally obtained from an RRAM in-memory module.

25 Summary

Emerging devices such as RRAM offers different options to improve computing efficiency:

- **Bring memory as close to logic as possible, still largely based on conventional architecture**

- **Neuromorphic computing in artificial neural networks**

- **More bio-inspired, taking advantage of the internal ionic dynamics at different time scales**

 Toward a general in-memory computing fabric based on a common physical substrate

R RAM is an interesting technology that is at the verge of being adopted in large scale. The benefits of RRAM can be realized in different approaches – embedded memory, artificial neural networks, and dynamic networks, with efficiencies improving as the device takes on more roles. The final goal may be to build a general, reconfigurable in-memory computing platform that is suitable for a broad range of data intensive tasks.

26 Acknowledgments

Grad students:
*Sung-Hyun Jo, *Kuk-Hwan Kim
*Siddharth Gaba, *Ting Chang
*Patrick Sheridan, *ShinHyun
Choi, *Jiantao Zhou, *Chao Du,
Jihang Lee, Wen Ma, Fuxi Cai,
Yeonjoo Jeong, Jong Hong Shin,
John Moon, Billy Schell, Qiwen
Wang *Eric Dattoli
*Wayne Fung, Lin Chen
*Seok-Youl Choi, *Woo Hyung Lee

PostDocs:
Dr. Mohammed Zidan,
Dr. Xiaojian Zhu
*Dr. Yuchao Yang
*Dr. Sungho Kim
*Dr. Bing Chen
*Dr. Taeho Moon
*Dr. Zhongqing Ji
* Dr. Qing Wan

Visiting scholars:
*Dr. L. Liu
*Xiaojie Hao

* alumni

•Prof. Z. Zhang, Prof. M. Flynn, UM
•Dr. G. Kenyon, LANL
•Prof. C. Teuscher, PSU
•Prof. D. Strukov, UCSB,
•Prof. J. Hasler, GeorgiaTech,
•Prof. R. Li, CAS, China
•Dr. I. Valov, Prof. R. Waser

Funding:
•ADA center through the DARPA/SRC JUMP program, DARPA UPSIDE program, DARPA ACCESS program
•National Science Foundation (ECS-0601478, CCF-0621823, ECCS-0804863, CNS-0949667, ECCS-0954621).
•DARPA SyNAPSE program
•Air Force MUTranslationalRI program, Air Force q-2DEG program
•Engineering Research (ETR) Grant

S tudies reported in this work will not be possible without the dedicated and creative work by students and postdocs, productive collaborations, and generous support from funding agencies.

About
the Editors

Rajiv Joshi

Dr. Rajiv V. Joshi is a research staff member and key technical lead at T. J. Watson research center, IBM. He received his B.Tech I.I.T (Bombay, India), M.S (M.I.T) and Dr. Eng. Sc. (Columbia University). His novel interconnects processes and structures for aluminum, tungsten and copper technologies which are widely used in IBM for various technologies from sub-0.5µm to 14nm. He has led successfully predictive failure analytic techniques for yield prediction and also the technology-driven SRAM at IBM Server Group. He has extensively worked on novel memory designs. He commercialized these techniques. He received 3 Outstanding Technical Achievement (OTAs), 3 highest Corporate Patent Portfolio awards for licensing contributions, holds 60 invention plateaus and has over 235 US patents and over 354 including international patents. His interests are in in-memory computation, CNN, DNN accelerators and Quantum computing. He has authored and co-authored over 200 papers. He has given over 45 invited/keynote talks and given several Seminars. He is awarded prestigious IEEE Daniel Noble award for 2018. He received the Best Editor Award from IEEE TVLSI journal. He is recipient of 2015 BMM award. He is inducted into New Jersey Inventor Hall of Fame in Aug 2014 along with pioneer Nicola Tesla. He is a recipient of 2013 IEEE CAS Industrial Pioneer award and 2013 Mehboob Khan Award from Semiconductor Research Corporation. He is a member of IBM Academy of technology and a master inventor. He served as a Distinguished Lecturer for IFEE CAS and EDS society. He is currently Distinguished Lecturer for CEDA. He is IEEE, ISQED and World Technology Network fellow and distinguished alumnus of IIT Bombay. He serves in the Board of Governors for IEEE CAS as industrial liaison. He serves as an Associate Editor of TVLSI. He will and has served on committees of DAC 2019, AICAS 2019, ISCAS, ISLPED (Int. Symposium Low Power Electronic Design), IEEE VLSI design, IEEE CICC, IEEE Int. SOI conference, ISQED and Advanced Metallization Program committees. He initiated IBM CAS EDS symposium at IBM in 2017 and will continue into 2018 with Artificial Intelligence as the focal area. He served as a general chair for IEEE ISLPED. He is an industry liaison for universities as a part of the Semiconductor Research Corporation. Also he is in the industry liaison committee for IEEE CAS society.

Matthew Ziegler

Dr. Matthew M. Ziegler is a Principal Research Staff Member at the IBM T. J. Watson Research Center, Yorktown Heights, NY. He received the Ph.D. degree in electrical engineering from the University of Virginia, Charlottesville, in 2004. Since joining IBM Research in 2004, he received several technical accomplishment awards in the areas of processor design, design automation, and low power design. Dr. Ziegler has directly participated in the design of IBM's Power Systems, z Systems, and BlueGene families of products. His research has recently focused on AI accelerator design, machine learning for CAD, VLSI design productivity, optimization, and low power design. This work has led to design methodologies and design automation systems used throughout IBM. He is a recipient of the 2018 Mehboob Khan Award from the Semiconductor Research Corporation and is a member of the IBM Academy of Technology. He has served on various conference committees, including being a general chair for ISLPED. He has recently served as a TPC chair for the 2018 and 2019 AI Compute Symposiums.

Arvind Kumar

Arvind Kumar is a Research Staff Member and Manager at the IBM T.J. Watson Research Center in Yorktown Heights, NY. He has worked extensively on device design, characterization, and modeling for several IBM technologies. He has recently shifted his research interests to AI and heterogeneous integration. In addition to numerous invited talks and panel presentations, he has been involved with the IEEE Rebooting Computing initiative, serving as General Chair of ICRC 2017. He currently manages a team focused on next generation AI hardware. He holds a PhD in Electrical Engineering and Computer Science from MIT.

Eduard Alarcón

Prof. Eduard Alarcón received the M. Sc. (National award) and Ph.D. degrees (honors) in Electrical Engineering from the Technical University of Catalunya (UPC BarcelonaTech), Spain, in 1995 and 2000, respectively. Since 1995 he has been with the Department of Electronics Engineering at the School of Telecommunications at UPC, where he became Associate Professor in 2000. From August 2003 to January 2004, July-August 2006 and July-August 2010 he was a Visiting Professor at the CoPEC center, University of Colorado at Boulder, US, and during January-June 2011 he was Visiting Professor at the School of ICT/Integrated Devices and Circuits, Royal Institute of Technology (KTH), Stockholm, Sweden. During the period 2006-2009 he was Associate Dean of International Affairs at the School of Telecommunications Engineering, UPC. He has co-authored more than 400 scientific publications, 7 books, 8 book chapters and 12 patents, and has been involved in different National, European (H2020 FET-Open, Flag-ERA) and US (DARPA, NSF) R&D projects within his research interests including the areas of on-chip energy management and RF circuits, energy harvesting and wireless energy transfer, nanosatellites, and nanotechnology-enabled wireless communications. He has received the Google Faculty Research Award (2013), Samsung Advanced Institute of Technology Global Research Program gift (2012), and Intel Honor Programme Fellowship (2014). He has given 30 invited, keynote and plenary lectures and tutorials in Europe, America, Asia and Oceania, was appointed by the IEEE CAS society as distinguished lecturer for 2009-2010 and lectures yearly MEAD courses at EPFL. He is elected member of the IEEE CAS Board of Governors (2010-2013), member of the IEEE CAS long term strategy committee, Vice President Finance of IEEE CAS (2015) and Vice President for Technical Activities of IEEE CAS (2016-2017, and 2018-2019). He was recipient of the Myril B. Reed Best Paper Award at the 1998 IEEE Midwest Symposium on Circuits and Systems. He was the invited co-editor of a special issue of the Analog Integrated

Circuits and Signal Processing journal devoted to current-mode circuit techniques, a special issue of the International Journal on Circuit Theory and Applications, invited associate editor for a IEEE TPELS special issue on PwrSOC. He co-organized special sessions related to on-chip power management at IEEE ISCAS03, IEEE ISCAS06 and NOLTA 2012, and lectured tutorials at IEEE ISCAS09, ESSCIRC 2011, IEEE VLSI-DAT 2012 and APCCAS 2012. He was the 2007 Chair of the IEEE Circuits and Systems Society Technical Committee on Power Circuits. He is acting as general co-chair of DCIS 2017, Barcelona and IEEE ISCAS 2020, Seville. He was the General co-chair of the 2014 international CDIO conference, the technical program co-chair of the 2007 European Conference on Circuit Theory and Design - ECCTD07 and of LASCAS 2013, Special Sessions co-chair at IEEE ISCAS 2013. He served as an Associate Editor of the IEEE Transactions on Circuits and Systems - II: Express briefs (2006-2007) and Associate Editor of the Transactions on Circuits and Systems – I: Regular papers (2006-2012) and currently serves as Associate Editor Elsevier's Nano Communication Networks journal (2009-), Journal of Low Power Electronics (JOLPE) (2011-) and in the Senior founding Editorial Board of the IEEE Journal on IEEE Journal on Emerging topics in Circuits and Systems, of which he is currently Editor-in-Chief (2018 and 2019).

About
the Authors

Rob Aitken

Rob Aitken is an ARM Fellow and technology lead for ARM Research. He is responsible for technology direction of ARM research, including identifying disruptive technologies and monitoring the global technology landscape. His research interests include emerging technologies, resilient computing, and statistical design. He has published over 80 technical papers, on a wide range of topics including impacts of technology scaling, statistics of memory bit cell variability and reliability. He holds 30 US patents. Dr. Aitken joined ARM as part of its acquisition of Artisan Components in 2004. He has given keynote addresses, tutorials and short courses at conferences and universities worldwide. He holds a Ph.D. from McGill University in Canada. Dr. Aitken is an IEEE Fellow, and serves on a number of conference and workshop committees.

Lisa Amini

Dr. Lisa Amini is the Director of IBM Research Cambridge, which includes the newly announced MIT-IBM Watson AI Lab. The MIT-IBM Watson AI Lab is dedicated to fundamental artificial intelligence (AI) research with the goal of propelling scientific breakthroughs in four research pillars: AI Algorithms, the Physics of AI, the Application of AI to industries, and Advancing shared prosperity through AI; all of which leverage and pioneer machine learning, deep learning, and machine reasoning algorithms. Lisa was previously Director of Knowledge & Reasoning Research in the Cognitive Computing group at IBM's TJ Watson Research Center in New York, and she is also an IBM Distinguished Engineer.

Lisa was the founding Director of IBM Research Ireland, and the first woman Lab Director for an IBM Research Global (i.e., non-US) Lab (2010-2013). In this role she developed the strategy and led researchers in advancing science and technology for intelligent urban and environmental systems (Smarter Cities), with a focus on creating analytics, optimizations, and systems for sustainable energy, constrained resources (e.g., urban water management), transportation, and the linked open data systems that assimilate and share data and models for these domains.

Previously, Lisa was Senior Manager of the Exploratory Stream Processing Research Group at the IBM TJ Watson Research Center. She was the founding Chief Architect for IBM's InfoSphere Streams product. The Streams product is the result of a Research technology, System S, for which Lisa was also architectural lead from inception. Streams is a software platform for continuous, high throughput, and low latency mining of intelligence from massive amounts of sensor and other machine generated data. She also led her team in formative Smarter Planet/ Cities pilots analyzing real-time data for cyber security, manufacturing, telecom, market data analysis, radio astronomy, environmental (water) monitoring, and transportation.

Lisa has served on program committees, hosted panels, and presented keynotes and papers in numerous IEEE, ACM and other conferences and workshops. She has worked at IBM the areas of AI and Cognitive Computing, Smarter Cities, Stream Processing, Distributed and high performance systems, Content Distribution, Multimedia, and Networking for over 25 years. She earned her PhD degree in Computer Science from Columbia University.

Andreas G. Andreou

Andreas G. Andreou is a professor of electrical and computer engineering, computer science and the Whitaker Biomedical Engineering Institute, at Johns Hopkins University. Andreou is the co-founder of the Johns Hopkins University Center for Language and Speech Processing. Research in the Andreou lab is aimed at brain inspired microsystems for sensory information and human language processing. Notable microsystems achievements over the last 25 years, include a contrast sensitive silicon retina, the first CMOS polarization sensitive imager, silicon rods in standard foundry CMOS for single photon detection, hybrid silicon/silicone chip-scale incubator, and a large scale mixed analog/digital associative processor for character recognition. Significant algorithmic research contributions for speech recognition include the vocal tract normalization technique and heteroscedastic linear discriminant analysis, a derivation and generalization of Fisher discriminants in the maximum likelihood framework. In 1996 Andreou was elected as an IEEE Fellow, "for his contribution in energy efficient sensory Microsystems."

Jeff Burns

Jeff Burns is the Director of AI Compute at IBM Research. He received his B.S. in Engineering from UCLA, and his M.S. and Ph.D. in Electrical Engineering from U.C. Berkeley. In 1988 he joined the IBM T.J. Watson Research Center and worked in layout automation and processor design. In 1996 he joined the IBM Austin Research Lab where he worked on the first 1 GHz PowerPC; he then managed the Exploratory VLSI Design group. In 2003 he returned to Watson to work on IBM Research's annual study into the future of IT. He then managed a program exploring a streaming-oriented supercomputer, followed the VLSI Design department, focusing on high-end processors, SoC designs, and 3D. He has held prior roles as the Director of VLSI Systems as well as the Director of Systems Architecture and Design, managing the Division's activities in VLSI design, design automation, microprocessor and systems architecture, and accelerator design.

Mike Davies

Mike Davies leads Intel's Neuromorphic Computing Lab. Since joining Intel Labs in 2014, he has researched neuromorphic prototype architectures and is responsible for Intel's recently announced Loihi research chip. Previously, as a founding employee of Fulcrum Microsystems and its director of silicon engineering, Mike pioneered high performance asynchronous design methodologies as applied to several generations of industry-leading Ethernet switch products. He joined Intel in 2011 by Intel's acquisition of Fulcrum.

Todd Hylton

Dr. Todd Hylton is the Executive Director of the Contextual Robotics Institute and Professor of Practice in the Electrical and Computer Engineering Department at UC San Diego. His research interests include novel computing systems and their application to autonomous vehicle and robotic systems. Prior to his appointment at UC San Diego, he was Executive Vice President of Strategy and Research at Brain Corporation, a San Diego-based robotics startup. From 2007 to 2012, Dr. Hylton served as a Program Manager at DARPA where he started and managed a number of projects including the Nano Air Vehicle program, the SyNAPSE program and the Physical Intelligence program. Prior to DARPA, he ran a nanotechnology research group at SAIC, co-founded 4Wave, a specialty semiconductor equipment business, and served as CTO of Commonwealth Scientific Corporation. Dr. Hylton received his Ph.D. in Applied Physics from Stanford University in 1991 and his B.S. in Physics from M.I.T. in 1983.

Wei D. Lu

Wei D. Lu is a Professor in the Electrical Engineering and Computer Science department at the University of Michigan, and Director of the Lurie Nanofabrication Facility. He received B.S. in physics from Tsinghua University, Beijing, China, in 1996, and Ph.D. in physics from Rice University, Houston, TX in 2003. From 2003 to 2005, he was a postdoctoral research fellow at Harvard University, Cambridge, MA. He joined the faculty of the University of Michigan in 2005. His research interest includes resistive-random access memory (RRAM), memristor-based logic circuits, neuromorphic computing systems, aggressively scaled transistor devices, and electrical transport in low-dimensional systems. To date Prof. Lu has published over 100 journal articles with 20,000 citations and h-factor of 61. He is an IEEE Fellow, a recipient of the NSF CAREER award, and co-founder and Chief Scientist of Crossbar, Inc.

Jan M. Rabaey

Prof. Rabaey holds the Donald O. Pederson Distinguished Professorship at the University of California at Berkeley. He is a founding director of the Berkeley Wireless Research Center (BWRC) and the Berkeley Ubiquitous SwarmLab.

Prof. Rabaey has made high-impact contributions to a number of fields, including advanced wireless systems, low power integrated circuits, sensor networks, and ubiquitous computing. His current interests include the exploration of the interaction between the cyber and the biological worlds.

He is the recipient of major awards, amongst which the IEEE Mac Van Valkenburg Award, the European Design Automation Association (EDAA) Lifetime Achievement award, the Semiconductor Industry Association (SIA) University Researcher Award, and the SRC Aristotle Award. He is an IEEE Fellow and a member of the Royal Flemish Academy of Sciences and Arts of Belgium. He has been involved in a broad variety of start-up ventures.

Naveen Verma

Naveen Verma received the B.A.Sc. degree in Electrical and Computer Engineering from the UBC, Vancouver, Canada in 2003, and the M.S. and Ph.D. degrees in Electrical Engineering from MIT in 2005 and 2009 respectively. Since July 2009 he has been with the Department of Electrical Engineering at Princeton University. His research focuses on advanced sensing systems, exploring how systems for learning, inference, and action planning can be enhanced by algorithms that exploit new sensing and computing technologies. This includes research on large-area, flexible sensors, energy-efficient statistical-computing architectures and circuits, and machine-learning and statistical-signal-processing algorithms. Prof. Verma has served as a Distinguished Lecturer of the IEEE Solid-State Circuits Society, and currently serves on the technical program committees for ISSCC, VLSI Symp., DATE, and IEEE Signal-Processing Society (DISPS). Prof. Verma is recipient or co-recipient of a number of awards for his research and teaching.